Economics of Research and Innovation in Agriculture

NBER

NATIONAL BUREAU *of* ECONOMIC RESEARCH

A National Bureau of Economic Research Conference Report

NBER
NATIONAL BUREAU *of* ECONOMIC RESEARCH

Economics of Research and Innovation in Agriculture

Edited by **Petra Moser**

The University of Chicago Press

Chicago and London

The University of Chicago Press, Chicago 60637
The University of Chicago Press, Ltd., London

Published 2021
Printed in the United States of America

30 29 28 27 26 25 24 23 22 21   1 2 3 4 5
ISBN-13: 978-0-226-77905-8 (cloth)
ISBN-13: 978-0-226-77919-5 (e-book)
DOI: https://doi.org/10.7208/chicago/9780226779195.001.0001

Library of Congress Cataloging-in-Publication Data

Names: Moser, Petra, editor. | National Bureau of Economic Research, organizer.
Title: Economics of research and innovation in agriculture / Petra Moser, editor.
Other titles: NBER conference report.
Description: Chicago ; London : The University of Chicago Press, 2021. | Series: A National bureau of economic research conference report | Papers from an NBER conference of the same name, held in Washington, DC, on 17 May 2019. | Includes bibliographical references and index.
Identifiers: LCCN 2021009480 | ISBN 9780226779058 (cloth) | ISBN 9780226779195 (ebook)
Subjects: LCSH: Agriculture—Research—United States—Congresses. | Research—United States—Finance—Congresses. | Agricultural innovations—United States—Congresses. | Agricultural innovations—Economic aspects—United States—Congresses. | Research and development projects—Economic aspects—United States—Congresses. | LCGFT: Conference papers and proceedings.
Classification: LCC S541 .E35 2021 | DDC 338.1072—dc23
LC record available at https://lccn.loc.gov/2021009480

♾ This paper meets the requirements of ANSI/NISO Z39.48–1992 (Permanence of Paper).

Relation of the Directors to the Work and Publications of the NBER

1. The object of the NBER is to ascertain and present to the economics profession, and to the public more generally, important economic facts and their interpretation in a scientific manner without policy recommendations. The Board of Directors is charged with the responsibility of ensuring that the work of the NBER is carried on in strict conformity with this object.

2. The President shall establish an internal review process to ensure that book manuscripts proposed for publication DO NOT contain policy recommendations. This shall apply both to the proceedings of conferences and to manuscripts by a single author or by one or more co-authors but shall not apply to authors of comments at NBER conferences who are not NBER affiliates.

3. No book manuscript reporting research shall be published by the NBER until the President has sent to each member of the Board a notice that a manuscript is recommended for publication and that in the President's opinion it is suitable for publication in accordance with the above principles of the NBER. Such notification will include a table of contents and an abstract or summary of the manuscript's content, a list of contributors if applicable, and a response form for use by Directors who desire a copy of the manuscript for review. Each manuscript shall contain a summary drawing attention to the nature and treatment of the problem studied and the main conclusions reached.

4. No volume shall be published until forty-five days have elapsed from the above notification of intention to publish it. During this period a copy shall be sent to any Director requesting it, and if any Director objects to publication on the grounds that the manuscript contains policy recommendations, the objection will be presented to the author(s) or editor(s). In case of dispute, all members of the Board shall be notified, and the President shall appoint an ad hoc committee of the Board to decide the matter; thirty days additional shall be granted for this purpose.

5. The President shall present annually to the Board a report describing the internal manuscript review process, any objections made by Directors before publication or by anyone after publication, any disputes about such matters, and how they were handled.

6. Publications of the NBER issued for informational purposes concerning the work of the Bureau, or issued to inform the public of the activities at the Bureau, including but not limited to the NBER Digest and Reporter, shall be consistent with the object stated in paragraph 1. They shall contain a specific disclaimer noting that they have not passed through the review procedures required in this resolution. The Executive Committee of the Board is charged with the review of all such publications from time to time.

7. NBER working papers and manuscripts distributed on the Bureau's web site are not deemed to be publications for the purpose of this resolution, but they shall be consistent with the object stated in paragraph 1. Working papers shall contain a specific disclaimer noting that they have not passed through the review procedures required in this resolution. The NBER's web site shall contain a similar disclaimer. The President shall establish an internal review process to ensure that the working papers and the web site do not contain policy recommendations, and shall report annually to the Board on this process and any concerns raised in connection with it.

8. Unless otherwise determined by the Board or exempted by the terms of paragraphs 6 and 7, a copy of this resolution shall be printed in each NBER publication as described in paragraph 2 above.

Contents

Introduction

Petra Moser

Over the last 50 years, mechanical, biological, and chemical innovations have more than doubled agricultural output while scarcely changing input quantities (Alston et al. 2010). In 1957, Zvi Griliches estimated that the internal rate of return (IRR) for research on new corn hybrids was around 40 percent. A meta-analysis of research and development (R&D) productivity estimates for 1965 to 2005 suggests even higher returns for those years, with a median estimate of 45 percent (Fuglie and Heisey 2007).

Yet returns to agricultural R&D are exceedingly difficult to measure. Even when costs and benefits are known, creating accurate summary statistics can be challenging. For example, an analysis of 2,242 investment evaluations between 1958 and 2011 has found that calculating a modified internal rate of return instead of the standard IRR is associated with an enormous decline in reported returns to agricultural R&D, reducing the estimated median annual return from 39 percent to less than 10 percent (Hurley, Rao, and Pardey 2014).[1]

Petra Moser is a professor of economics at New York University, a research fellow of the Center for Economic Policy Research, and a research associate of the National Bureau of Economic Research.

Early drafts of the chapters in this book were presented and discussed at the NBER conference on the "Economics of Research and Innovation in Agriculture," Washington, DC, May 17, 2019, funded by the US Department of Agriculture. I am grateful for the excellent comments on this chapter from Matt Clancy, James McDonald, Paul Rhode, Michael Roberts, Brian Wright, and two anonymous readers. For acknowledgments, sources of research support, and disclosure of the author's material financial relationships, if any, please see https://www.nber.org/books-and-chapters/economics-research-and-innovation-agriculture/introduction-economics-research-and-innovation-agriculture.

1. Another potential issue is that some of the welfare benefits of agricultural innovation may accrue to consumers in the form of lower prices for agricultural goods. Low price elasticities of demand for agricultural products imply that productivity gains from freely accessible agricul-

Moreover, many recent studies find that returns to agricultural research have been declining of late. Andersen, Alston, Pardey, and Smith (2018) document that US multifactor farm productivity grew at an annual average rate of 1.16 percent per year during 1990–2007, down from 1.42 percent per year for 1910–2007. They also find that US yields of major crops grew at an annual average rate of 1.17 percent for 1990–2009 compared with 1.81 percent for 1936–90. Similarly, an analysis of research inputs and total factor productivity (TFP) between 1970 and 2007 indicates that TFP growth declined slightly in agriculture, while effective research investments rose by a factor of two (Bloom et al. 2019), suggesting that research productivity declined by a factor of nearly four, equivalent to an average decline of 3.7 percent per year.

Intensifying the potential threat of diminished productivity, the share of gross domestic product (GDP) to agricultural R&D has declined in many wealthy countries. Historically, the US public sector has been a top performer in worldwide agricultural R&D. This situation, however, has changed significantly in recent years, and the United States has lost its dominant position, falling behind China in 2009 through at least 2013 (Clancy, Fuglie, and Heisey 2016). In 1995, total global spending on agricultural R&D was around $33 billion. Roughly two-thirds of this spending originated from governments, universities, and nonprofits, while one-third originated from profit-motivated R&D (Pardey and Beintema 2001). Five years later, by 2000, total global spending was roughly the same, but the share of public to profit-motivated R&D had changed to 60 and 40 percent (Pardey et al. 2006), highlighting a growing reliance on industry funding for agricultural R&D.

This book provides new evidence on the potential impact of this shift from public to private sector funding and, more generally, furthers our understanding of the returns to public and private spending R&D. Measuring research and innovation is difficult in any field, but particularly in agriculture, and data constraints create major challenges for empirical analyses. To address these challenges, chapters in this book present original data sets ranging from text-based measures of innovation to animal-level data on dairy cow performance and fine-grained data on yields. Comments on these chapters discuss remaining measurement challenges and suggest promising directions for future data efforts and analyses.

Thematically, the chapters examine the sources of agricultural knowledge and investigate challenges for measuring the returns to the adoption of new agricultural technologies, survey knowledge spillovers from universities to agricultural innovation, and explore interactions between university engage-

tural innovations reduce the price of agricultural goods (Guttman 1978), making consumers the primary beneficiaries of such innovations. With free trade and reasonable transport costs, these welfare gains diffuse across domestic and foreign consumers, reducing domestic consumers' willingness to pay.

ment and scientific productivity. Analyses of agricultural venture capital point to that industry as an evolving source of funding for agricultural R&D.

Methodologically, the research in this book spans a diverse spectrum, from archival research and text analysis to survey design and structural estimates. Yet all these individual contributions share some common traits. Several chapters use more fine-grained data than have been previously available to challenge prior findings (e.g., chapters 2 and 4) or resolve unanswered questions (e.g., chapter 3). Individual chapters use novel empirical methods to understand the sources of agricultural innovation (chapter 1), while others provide descriptions of important and new phenomena that are important for agricultural innovation (chapters 5 and 6). Chapters with a historical focus provide important insights that speak to our current challenges, such as agricultural adaptation to climate change. Building on this work, discussions for each chapter outline promising directions for future research.

I.1 Tracing Agricultural Productivity to Its Source

In their chapter, "The Roots of Agricultural Innovation: Patent Evidence of Knowledge Spillovers," Matt Clancy, Paul Heisey, Yongjie Ji, and GianCarlo Moschini investigate knowledge spillovers from innovations *outside of agriculture* as sources of agricultural innovation. While many previous analyses have investigated knowledge spillovers, nearly all these studies have focused on spillover between different segments of agricultural R&D (e.g., Evenson 1989) or across states or countries (Alston 2002). This chapter extends prior studies in two major directions by (1) examining spillovers from other industries into agriculture and (2) introducing a new method to measure knowledge spillovers through text analysis.

Using the full text of US agricultural patents issued between 1976 and 2016, Clancy and his coauthors construct three complementary measures of knowledge spillovers: (1) citations to nonagricultural patents, (2) citations to scientific publications in nonagricultural journals, and (3) a text-analysis algorithm that identifies "text-novel concepts" that are novel to agricultural patents but not to other technology fields. The authors apply these three measures to patents in subsectors of agriculture: animal health, biocides, fertilizer, machinery, plants, and research tools.

Analyses of all three measures indicate that more than half of all patents in agriculture have benefitted from knowledge sources outside of agriculture (figure I.1). In three of the six subsectors—animal health, fertilizer, and machinery—more than half of all spillovers into agriculture appear to have originated from other industries. In animal health, the share of outside knowledge among cited patents is extremely large, on the order of 90 percent. In only one subsector—plants—knowledge flows typically originate from agricultural R&D.

Nonagricultural sources of knowledge flows into agriculture are, however, rarely completely detached from agricultural research. For example, agricultural patents are more likely to cite scientific publications in biology and chemistry compared with publications in other journals. Agricultural patents are more likely to cite or share text-novel concepts with the nonagricultural patents of firms that have at least one agricultural patent in their portfolio.

The new text-analysis measure of spillovers is a major contribution of this chapter, and it introduces a useful complement to citations as a measure of knowledge flows. Methodologically, Clancy and his coauthors define text-novel concepts as words and phrases (strings) that are new in agricultural patents in the second half of their data (for patents with application years between 1996 and 2018). First, they identify roughly 100 text-novel concepts in each of the six subsectors. Then they search all US patents in other sectors (outside of their six subsectors) for prior mentions of these concepts. For example, the string *pyrimethamine* does not appear in any animal health patents before 1996 but is a common term in animal health patents afterward, making it a text-novel concept. When earlier patents on human health mention pyrimethamine, their measure records an incidence of knowledge spillover from human health to animal health.

Using these new text-based measures, the authors make two important points. First, they show that knowledge spillovers from nonagricultural sources are essential to agricultural innovation. Second, they find that citation-based measures of knowledge spillovers, which have been used as the standard measure of knowledge spillovers, overstate the share of knowledge spillovers *within* agriculture relative to text-based measures (figure I.1). Within the agricultural sector, the authors identify several areas in which findings from citation-based measures may be misleading. In biocides, for example, most patents cite nonagricultural patents and journals, which suggests that most spillovers originate from other disciplines. Using the measure of text-novel concepts, however, the authors show that these concepts are never mentioned in earlier patents outside of biocides, which indicates that they may have originated in biocides.

Their discussant, Alberto Galasso, emphasizes that these findings have important implications for our understanding of how shocks propagate through the economy through industry linkages (Barrot and Sauvagnat 2016). He also suggests a potential refinement for estimates of knowledge spillovers by controlling for the size of technology fields. A relatively small field like animal health may appear to draw more knowledge from a large field, like chemistry, simply because chemistry is a very large field; controlling for field size will address this issue. Galasso further highlights the importance of distinguishing involuntary spillovers from intentional knowledge transfer through licensing contracts between nonagricultural and agricultural firms. This concept is picked up and extended in later chapters on knowledge flows between universities and industry.

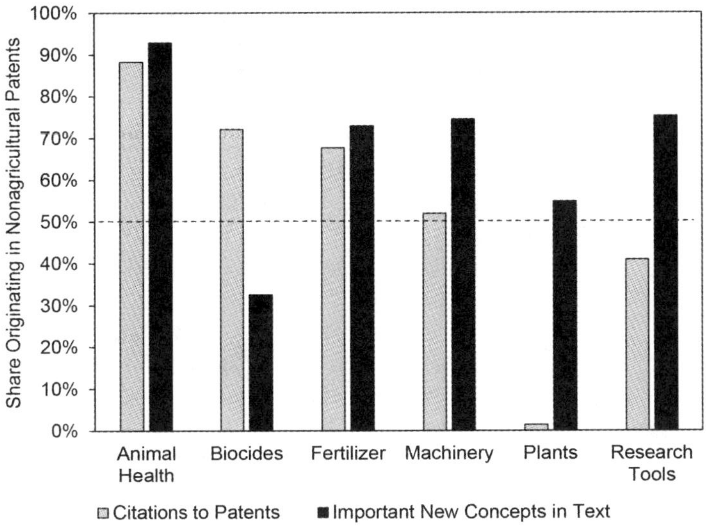

Fig. I.1 Knowledge spillovers into agriculture

Note: Knowledge spillovers into agricultural patents from other fields, measured through the traditional measure of *Citations to Patents* and the author's new text-based measure of *Important New Concepts in Text*. This latter variable captures concepts that do not appear in a given subsector before 1996 but become important afterward. The figure is based on data from chapter 1 in this book.

I.2 Selection as a Challenge for Measuring Returns to Biological Innovation

A chapter by Jared Hutchins, Brent Hueth, and Guilherme Rosa on "Quantifying Heterogeneous Returns to Genetic Selection: Evidence from Wisconsin Dairies" uses individual-level microdata on milk production in a structural model to estimate the impact of genetic selection. The dairy industry has experienced a 3 to 4 percent increase in milk yields per year; half of this increase has been attributed to genetic improvement in the quality of bulls. Yet the match between the bull and the dame (the mother of a new cow) may be just as important as the quality of the bull. Such selection is a common problem in estimating returns to agricultural innovation. For hybrid corn, for example, a substantive share of the increase in yields after the adoption of hybrid corn is due to the fit between the hybrid seed and its most productive environment, as Griliches (1957) has shown for the early 20th-century United States and Suri (2011) for modern-day Kenya.

Observing and identifying selection in the dairy industry, however, is difficult because success takes several years to observe. For corn, the success of a new match can be observed within the season. Cows, however, take three years to mature before they produce milk. This delay between the matching of a dame and a bull and the breeder's ability to observe the milk production of their offspring is simply too long to allow for experimental learning. As a result, genetic improvements in dairy occur gradually through an endogenous process of selection that is mediated by demand and supply.

Hutchins, Hueth, and Rosa estimate the contribution of this selection process using uniquely detailed data on the "genetic merit" of individual bulls from the Dairy Herd Improvement (DHI) program. Going back to 1908, this program of the US Department of Agriculture (USDA) covers roughly half of all dairy herds in the United States. Widely adopted since the early 1960s, artificial insemination technologies have created unprecedented opportunities to observe the performance of bulls, who can now produce thousands of offspring. Every daughter of a bull contributes new data, improving the estimates of milk production associated with his genes. The authors exploit these data to estimate a structural model of genetic improvement and selection in the form of assortative matching between a high-value cow and a bull.

Estimates from a structural model of returns to high-yield genetics imply that 75 percent of these returns are driven by selection in the form of assortative matching. Exploiting animal-level data, the authors show that productivity gains are driven by matching at the level of animals and not just at the farm. In other words, they show that productivity in dairy has increased not only because better farmers choose better bulls but also because farmers match productive cows with productive bulls.

These findings indicate that farmers are critical to determining the returns to biological innovation today. This is similar to the role they played in US innovation historically, when farmers often discovered new varieties of food and feed crops. Olmstead and Rhode (2008), for example, examine the challenges that informational problems and cross-fertilization created for innovations by private farmers and breeders in cotton. According to Robert Evenson, until the end of the 19th century, all crucial mechanical inventions in agriculture were the work of farmers and local blacksmiths rather than of large corporations (cited in Wright 2012, 1718).

I.3 Innovation as a Response to Environmental Shocks

Expanding on the theme of farmers' role in selecting the most productive technologies, a chapter by Keith Meyers and Paul W. Rhode examines farmers' decisions to adopt heat-resistant corn hybrids after a series of catastrophic droughts and harvest failures in the 1930s. In "Yield Performance of Corn under Heat Stress: A Comparison of Hybrid and Open-Pollinated Seeds during a Period of Technological Transformation, 1933–55," Meyers and Rhode use newly recovered data from the archives of Zvi Griliches to reexamine the diffusion of hybrid corn seeds immediately following the Dust Bowl (1930–36).

Hybridization, which creates a new variety by crossing two corn (so-called filial F1) varieties, provided a new method of developing higher-yielding and more resilient seeds. Compared with the traditional open-pollinated seeds (which are simply allowed to propagate in the fields), hybrids yield more corn and take less time to mature. They also have stronger roots and thicker stalks, which make them less susceptible to breaking in wind or rain; they are more resistant to disease; and they are more likely to survive a drought. Yet hybrid seeds also cost more than open-pollinated seeds (Olmstead and Rhode 2008), and farmers cannot save hybrid seeds from their harvest to plant in the following year because the offspring of saved seeds return to the characteristics of the parental varieties (instead of exhibiting the desirable traits of the purchased hybrid seed). As a result, farmers who switch to hybrid seeds must buy new seeds from the breeder every year instead of building their own supply. These trade-offs led to an uneven adoption of hybrid corn, which Meyers and Rhode reexamine in their chapter.

Griliches (1957) showed that expected improvements in hybrid yields drove the adoption of hybrid corn in the Corn Belt and the Great Plains. Yet, Meyers and Rhode note, Griliches may have overlooked a significant link between the adoption of hybrids and a period of devastating droughts and crop failures during the Dust Bowl years of 1934 and 1936. Narrative historical evidence suggests that corn farmers learned about the benefits of planting drought-resistant hybrids by observing neighbors' crops failing or surviving during these droughts. The late Richard Sutch (2011) argued that drought resistance became more salient to farmers as a result of climate shocks, and he highlighted the USDA's role in promoting hybrid seeds after the Dust Bowl.

In fact, hybrid corn gained its most substantial foothold in US agriculture in 1937, just one year after the catastrophic harvest failures of 1936 (figure I.2), and was planted on more than 40 percent of corn acreage in the most productive counties of Iowa and Illinois.

To investigate whether hybrids did in fact mediate the effects of weather shocks—in the form of extreme heat and drought—Meyers and Rhode have returned to Griliches's archives to construct fine-grained geographic data on hybrid corn adoption and yields, matched with historical data on droughts. While existing analyses rely on state-level data, this substantial effort of data collection allows Meyers and Rhode to examine adoption patterns at the level of crop reporting districts (CRDs), roughly the size of 10 neighboring counties. This analysis indicates corn breeding allowed the corn frontier to move farther north, into Canada. Focusing on heat tolerance as a measure for tolerance to droughts, Meyers and Rhode show that hybrid corn grown in Iowa from 1928 to 1942 did exhibit heat tolerance relative to open-pollinated varieties, consistent with the findings of Sutch (2011). These results, however, do not replicate in other states, and reduced temperature sensitivity does not appear when comparing hybrid and open-pollinated yields grown in other states. This latter finding supports Griliches's decision to ignore drought tolerance in his analysis of hybrid adoption.

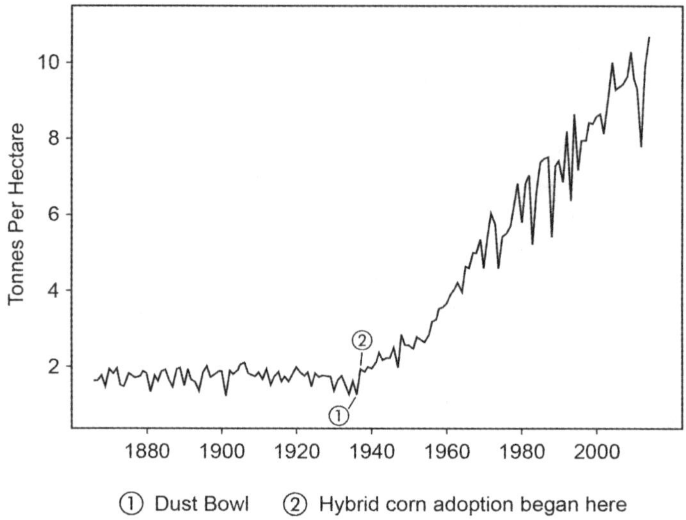

Fig. I.2 US corn yields, 1888–2014

Note: From Michael Robert's comment on the chapter by Meyers and Rhode in this book (see chapter 3), using data on corn yields from the USDA's National Agricultural Statistics Service (https://www.nass.usda.gov).

Their discussant, Michael Roberts, is even more skeptical than the authors of the view that the adoption of hybrid corn was a response to the Dust Bowl and issues a stark warning about the limits of technical change in agriculture as a response to climate change. Schlenker and Roberts (2009), for example, have shown that the number of extreme heat days above 29°C is the best predictor of corn yields. Modern data indicate that high-yielding genetically modified varieties that are prevalent today are even more sensitive to extreme heat than the traditional varieties (Lobell, Schlenker, and Costa-Roberts 2011).[2]

In the 20th century, US agriculture was able to capitalize on vast productivity gains by developing plants with immense yield potential (the maximum output given available sunlight and light) and by creating varieties to match the available sunlight and water across the United States while also processing massive amounts of nitrogen from fertilizers. Today, nitrogen is no longer a limiting factor, and the adoption of genetically modified crops (such as Roundup Ready corn) has made it easier to control weeds (Roundup, or glyphosate) and pests (through BT strains). Yet the large plants of today

2. Genetically engineered drought tolerance was introduced in corn hybrids in 2012 and became broadly available the following year. By 2016, 22 percent of total US planted corn acreage was drought tolerant. As the research of Richard Sutch as well as Meyers and Rhode would suggest, adoption has been concentrated in drought-prone regions (despite the hybrids' limited ability to protect against the most extreme droughts; McFadden et al. 2019).

with their deep roots require more water, leaving modern varieties vulnerable to droughts. The unusually hot summer of 2012 approached the temperatures of the Dust Bowl. Current climate models predict many more summers like 2012, with even hotter temperatures. Roberts warns that innovation in corn and other crops may be unable to deal with extreme temperatures. Plants have reached the biological limits of photosynthesis, requiring an entirely new approach for a second Green Revolution.

Recent advances in the emerging field of synthetic biology may offer a much-needed novel approach by targeting improvements in photosynthetic efficiency. For example, a survey article by Batista-Silva et al. (2020) discusses the progress and challenges of engineering improved photosynthesis through synthetic biology as a potential path toward improving the utilization of solar energy and carbon sources to produce food, fiber, and fuel.

I.4 Universities as a Source of Agricultural Innovation and Productivity Gains

Publicly funded research has been a major source of innovation and advances in agricultural productivity throughout American history (e.g., Shih and Wright 2011; Olmstead and Rhode 2008). Since their foundation under the Morrill Land-Grant Acts of 1862 (7 U.S.C. §301 et seq.), the original 52 land grant universities have been the key institutions in creating and disseminating agricultural innovations (Wright 2012), establishing vital links among universities, farmers, and industry. With the 1862 act, the US government allotted 30,000 acres of federal land per state to finance the foundation of practically oriented research and training universities.[3] The 1887 Hatch Act (7 U.S.C. § 361a et seq.) added research capabilities through state agricultural experiment stations, supported by grants of additional federal lands. In 1890, the second Morrill Act (7 U.S.C. §322 et seq.) increased the funding of these new colleges to $25,000 per year and specified that African Americans could receive education in existing land grant colleges and in new colleges designed for that purpose. Finally, in 1914, the Smith-Lever Act established a cooperative extension service to inform farmers about agricultural innovations and establish home instruction to help farmers learn about new agricultural techniques.

In its early decades of operation, the US land grant system supported agricultural productivity by encouraging the diffusion of European innovations. Evenson (1978), for example, documents that advances in agricultural productivity between 1870 and 1925 were strongly correlated with total real public spending on agricultural research during the preceding 18

3. Southern states had originally opposed the Morrill Act, and it only passed after the South seceded from the United States. As a result, none of the original 52 land grant colleges operated in the South.

years, but largely based on the adoption of European inventions. It took several decades, until the 1930s, for the system of land grant colleges and experiment stations to become an efficient source of domestic agricultural innovation (Huffman and Evenson 2006). Kantor and Whalley (2019) find that the establishment of agricultural experiment stations at existing land grant institutions through the Hatch Act of 1887 took between 20 and 30 years to increase land productivity in neighboring counties. Olmstead and Rhode (2002, 931–32) show that, with the exception of early advances in corn, yields for field crops only began to increase after 1930. US wheat yields increased only 1.75 bushel per acre between 1866 and 1939 but increased by about 2.25 percent per year afterward, doubling wheat yields by the 1970s.

Rosenberg and Nelson (1994) reason that the land grant college system was uniquely suited to resolve a fundamental tension created by industry funding for academic research. University research is typically "basic" research, aimed at understanding fundamentals, with payoffs that are often uncertain, distant, and exceedingly difficult to appropriate. By contrast, industry research targets specific problems and challenges with payoffs that are substantially more immediate and are expected to directly benefit the firm that funds the R&D. Due to this tension, many academics view industry funding as a direct threat to their research and academic integrity, as targeted problem-solving takes time from basic research and sometimes even threatens open communications that are critical to academic exchange. According to Rosenberg and Nelso, the institutional features of the land grant college, with a firm commitment to knowledge diffusion and the implementation of feedback from local users, are uniquely suited to easing the tension between basic and applied research, especially after the Smith-Lever Act of 1914 provided funding for agricultural extension.

In "Local Effects of Land Grant Colleges on Agricultural Innovation and Output," Michael J. Andrews estimates the effect of establishing a land grant college on invention and agricultural performance on surrounding locations. To make some progress toward identifying the causal effect of establishing a land grant college on invention, Andrews compares locations that received a land grant college to "runner-up" counties that competed for establishing a land grant college but ultimately lost. Comparing changes in patenting in college and runner-up counties, Andrews shows that patenting increased in winning countries (compared to runner-up counties) after the establishment of a land grant college.

Patents, however, are an extremely noisy and potentially biased measure of agricultural innovations. Agricultural innovations of a chemical or mechanical nature were patentable throughout this period, while seeds and other types of biological innovations had no intellectual property protection. Moreover, even among innovations that were patentable, there were large differences in the share of innovations that inventors chose to patent across sectors and over time. An analysis of innovations exhibited at world technology fairs between 1851 and 1915 shows that roughly half of all agricultural machinery was covered by patents throughout this period (Moser 2012). By contrast, chemical innovations were almost never patented at the beginning of this period and experienced a dramatic shift toward patenting after improvements in analytic methods reduced the effectiveness of secrecy as an alternative to patents. Biological innovations first became subject to intellectual property rights through the Plant Patent Act of 1930. Plant patents, however, are substantially narrower than utility patents, and they are limited to asexually reproducing plants (plants, such as apples and roses, that reproduce by roots, shoots, or buds). Plant patent protection excludes plants that reproduce sexually, through seeds, as well as potatoes and other tubers (Moser and Rhode 2012).[4]

To address these issues, Andrews uses historical data on the introduction of new wheat varieties from Clark, Martin, and Ball (1922) as an alternative, nonpatent measure of innovation. This measure shows that land grant counties were about five times more likely to introduce a new wheat variety compared with runner-up counties after the establishment of a land grant college.

These findings are consistent with earlier research by Olmstead and Rhode (2002) that documents how the land grant system helped create and diffuse critical innovations in wheat through the type of regional adaptive research for which the system had been designed. As the center of gravity of wheat production extended westward to less-favorable environments, breeders in the land grant system identified and selected varieties that could tolerate drought, cold, insect pests, rusts, and other fungal diseases in these newly established growing regions.

Investigating funding as a mechanism for encouraging innovation, Andrews shows that the effects of land grant colleges on local innovations were largest following the passage of legislation, such as the Hatch Act of 1887, which increased funding for agricultural research.

Turning to agricultural productivity, however, Andrews finds that compared with runner-up counties, land grant counties experienced only small (and often negligible) improvements in agricultural productivity, measured by improvements in yields, crop output, or the production of livestock. Andrews explains that the productivity benefits of land grant research may

4. Using new varieties of roses as a nonpatent measure of innovation, Moser and Rhode (2012) investigate whether the creation of plant patents in 1930 led to a significant increase in agricultural innovation. (Notably, most plant patents until the 1960s covered roses. Data on registrations of newly created roses indicate no increase in innovation after 1930: Less than 20 percent of new roses were patented, European breeders continued to create most new roses, and there was no increase in the number of new varieties per year after 1931. Instead, influential new varieties appear to have been a by-product of publicly funded research.)

have diffused beyond the borders of the college county through a combination of outreach and university engagement (as described in chapter 5 of this book).

Placing Andrews's results in the broader context of productivity spillovers suggests that the geographic diffusion of spillovers—beyond the county level—is a likely explanation for the weakness of county-level productivity effects. In a state-level analysis of productivity spillovers, Alston et al. (2010) show that over half of the measured within-state productivity gains result from public research investments made elsewhere. Alston et al. estimate that the average marginal internal rate of return of public research accruing within the source state is 18.9 percent, significantly less than the estimated overall IRR of 22.7 for the entire nation. Thus the "failure" of the land grant system may lie in Alston et al.'s focus on state-level agricultural priorities and a lack of specificity of their research to local (county-level) conditions rather than in low productivity gains overall.

In his discussion, Bhaven N. Sampat highlights the usefulness of this chapter for the broader literature on returns from publicly funded research, which has held up the land grant system as a model of technology transfer that was more successful than the current, post-Bayh-Dole system of patent licensing (Mowery et al. 2004). Sampat also reminds us of Brian Wright's (2012) positive assessment of the land grant system. Citing the findings of Olmstead and Rhode (2002), Wright (2012, 1719) reports that by 1919, more than three-quarters of US wheat acreage used new varieties that had not been developed before the Morrill Act.

Sampat also points out that a strict focus on the diffusion of specific varieties may miss the contributions of universities if academic research contributes research techniques and tools rather than new products. In the words of Griliches (1957, 502), "Hybrid corn was the invention of a method of inventing." Citing primarily Stackman, Bradfield, and Mangelsdorf (1967), Wright (2012, 1720–25) documents how research methods developed within the land grant system facilitated the development of new wheat varieties in Mexico after 1943 and supported research to improve rice in India and the Philippines. More recently, an analysis of drug development between 1985 and 2005 has shown that public sector labs *enable* two-thirds of marketed drugs, even though they only directly create one-tenth of these new drugs (Sampat and Lichtenberg 2011).

I.5 Industry Engagement and Scientific Productivity

In their research on "Academic Engagement, Commercialization, and Scholarship: Empirical Evidence from Agricultural and Life Scientists at US Land Grant Universities," Bradford Barham, Jeremy Foltz, and Ana Paula Melo examine links between industry funding and the activities, attitudes, and research choices of agricultural and life science faculty at land grant colleges. Their analysis focuses on two major questions: (1) What types of interactions are most likely to increase industry funding for faculty research? and (2) How does funding from industry influence the research of scientists? To answer the first question, the authors analyze two waves, conducted in 2005 and 2015, of a survey of faculty at all 52 original land grant colleges. To analyze interactions between faculty and industry, the authors distinguish academic *engagement* (in the form of sponsored research, collaborations, and presentations) from *commercialization* (which includes patenting, licensing, and start-ups).[5]

Survey responses from faculty at land grant colleges reveal that academic engagement has generated between 15 and 20 times more research funding than academic commercialization. Engagement dates back to the land grant universities' emphasis—since their inception in the 19th century—on practical agricultural and engineering sciences, formal extension appointments for faculty, and ongoing outreach with farms and firms to improve their performance. Dispelling the fear that engagement with industry crowds out research, the authors also find that faculty who are more engaged with industry publish more.

Notably, their surveys uncover important differences in the faculty-industry relations across universities (figure I.3), which suggests that the institutional characteristics of universities play an important role in shaping links between academia and industry. As universities have been affected by dwindling state and federal support (e.g., Ehrenberg 2012), understanding sources of funding becomes critical. In principle, the passage of the Bayh-Dole Act in 1981 has created a new framework to commercialize innovations and discoveries associated with federally sponsored research (Sampat 2006; Thursby and Thursby 2011). Yet the creation of stronger incentives at publicly funded institutions through Bayh-Dole appears to have failed to encourage innovation. The findings of Barham, Foltz, and Melo suggest that, at least for the agricultural sector, the key institutions for university-industry relations had already been established in the 19th century through the US system of land grant colleges.

Their discussant, Nicola Bianchi, emphasizes that this chapter is one of the most thorough analyses of university-industry relations to date but also proposes promising directions for future research. For example, Bianchi points out that there is room to investigate the links between declining government grants and faculty involvement in university-industry relations. Follow-on research could also take advantage of publicly available sources on research output, including patents and publications, to complement the chapter's rich existing data from faculty surveys.

5. This distinction is adopted to match recent papers on university-industry relations in Europe, such as Perkmann et al. (2013); Tartari, Perkmann, and Salter (2014); Tartari and Salter (2015); and Sengupta and Ray (2017).

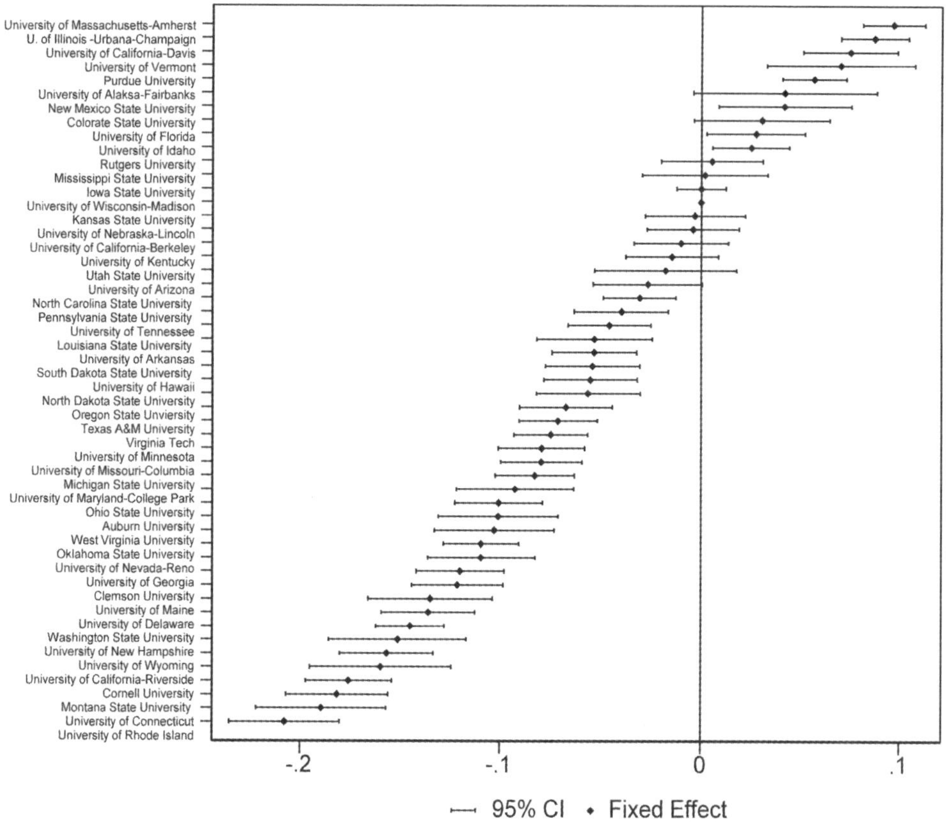

Fig. I.3 University-level probabilities of faculty engagement with industry

Note: From the chapter by Barham, Foltz, and Melo in this book (see chapter 5). OLS estimates and 95 percent confidence for 52 university fixed effects (with the University of Wisconsin-Madison as the excluded category). The dependent variable is an indicator that equals 1 if a faculty member is engaged in any type of university-industry relations (UIR). Estimates control for gender, being a professor, and having received a PhD from a land grant university. Standard errors are clustered at the university level.

I.6 Financing Future Innovations through Venture Capital

A final chapter on "Venture Capital and the Transformation of Private R&D for Agriculture" presents a forward-looking analysis of recent trends in the financing of innovations. In this chapter, Gregory D. Graff, Felipe de Figueiredo Silva, and David Zilberman document the dramatic expansion of venture capital (VC) investments in agriculture start-ups, especially in the wake of the financial crisis of 2008. Between the early 2000s and 2018, VC investments in start-ups focusing on agricultural R&D increased from just tens of millions to more than seven billion. Notably, VC investment in agriculture start-ups increased not only in absolute terms but also relative to the overall supply of capital invested by the public sector and by public firms. To perform their analysis, the authors combine data from three proprietary sources (Crunchbase, PitchBook, and VentureSource) to construct a new data set consisting of 4,500 start-ups in agriculture, with more than 10,000 financial transactions, including information on investments and exits.

Although, historically, private investment in agricultural R&D in emerging economies has been low (Pardey and Beintema 2001; Pardey et al. 2006), the authors report robust start-up activities in the larger emerging economies like India, China, and Brazil. In regression analyses, they examine potential causes for this shift, using data on 4,500 start-ups across 124 countries. Although the largest share of the start-ups in their sample operates in the United States (33 percent) and the European Union (23 percent), a significant share of the remaining 44 percent of start-ups is in emerging economies. The authors' regressions indicate that investments are strongly correlated with past liquidity events, suggesting that the expansion of VC investment in agriculture start-ups reflects a response to new investment opportunities in agriculture.

For a subset of these start-ups, their data also include information on investment and exit deals between 1981 and 2018. These data indicate that successful exits, in the forms of initial public offerings (IPOs) and mergers and acquisitions (M&A), led to higher VC investments. Comparing different types of exits, the authors find that prior IPOs are associated with a stronger increase in investments than prior M&As. These findings are important for researchers and policy makers who aim to support agricultural innovation and R&D. Overall, the authors conclude that venture capitalists' willingness to invest may have been affected by an increase in the ratio of agricultural prices to nonagricultural commodity prices, highly visible exits of major players in the agriculture technology space, changes in agricultural labor markets, and advances in enabling (general purpose) technologies, such as cheaper genome sequencing, genome editing, or increasing data capacity of sensors and networks.

A discussion by Michael Ewens suggests promising directions for future research. First, Ewens suggests extending the existing results with an in-depth analysis of a single source. Such an analysis would address empirical challenges that result from variation in the coverage of agricultural VCs across the merged sources. Some of these analyses may require hand-collecting additional data, especially to expand the coverage of agriculture start-ups in emerging economies. Second, Ewens recommends additional analyses *within* agriculture to identify areas that grew differentially after 2008, using data on agricultural prices. For example, a potential extension would apply an empirical strategy implemented by Ewens, Nanda, and Rhodes-Kropf (2018), which examines the effects of the cloud on VC in information technology. An extension to agriculture could exploit the effects of the same technology shock across different sectors within agriculture. Third, Ewens recommends examining the identities of investors, possibly by tracking the work histories of VC partners that choose to finance start-ups in agriculture. This question is particularly interesting and important because agriculture is a nontraditional investment for both VC and private equity.

I.7 Summing Up

Importantly, the economics of agricultural innovation is even broader than the research included in this volume. While this book is focused primarily on agricultural innovation in the United States, a rich literature in development economics examines forces that drive the adoption of agricultural innovations (e.g., Foster and Rosenzweig 1995; Conley and Udry 2010; Suri 2011)

Other recent research has examined the effects of restrictions on the supply of farm labor on agricultural innovation, using historical restrictions on immigration as a source of exogenous variation (Clemens, Lewis, and Postel 2018; San 2020). These papers build on a long tradition of economic research on endogenous technical change reaching back to Hicks (1932). In fact, much of what we know about endogenous technical change has been learned in the context of labor-saving innovations in agriculture (e.g., Hayami and Ruttan 1970). These analyses range from the adoption of tractors in the first half of the 20th century to co-robots (machines that work alongside humans) that weed crops today and grafting robots that replace humans in the labor-intensive task of grafting herbaceous seedlings of fruits and vegetable crops (Gallardo and Sauer 2018).

Despite these omissions, the chapters in this book outline diverse research that improves our understanding of agricultural innovation. This agenda spans several fields within economics, reaching from agricultural economics and economic history to finance and industrial organization. Authors of chapters, and their discussants, suggest promising opportunities for future research on the economics of agricultural innovation.

References

Alston, Julian M. 2002. "Spillovers." *Australian Journal of Agricultural and Resource Economics* 46 (3): 315–46.

Alston, Julian M., Matthew A. Andersen, Jennifer S. James, and Philip G. Pardey. 2010. *Persistence Pays: U.S. Agricultural Productivity Growth and the Benefits from Public R&D Spending*. New York: Springer.

Andersen, Matthew, Julian Alston, Philip Pardey, and Aaron Smith. 2018. "A Century of U.S. Farm Productivity Growth: A Surge Then a Slowdown." *Ameri-*

can Journal of Agricultural Economics 100:1072–90. https://doi.org/10.1093/ajae/aay023.

Barrot, J., and J. Sauvagnat. 2016. "Input Specificity and the Propagation of Idiosyncratic Shocks in Production Networks." *Quarterly Journal of Economics* 131:1543–92.

Batista-Silva, Willian, Paula da Fonseca-Pereira, Auxiliadora Oliveira Martins, Agustín Zsögön, Adriano Nunes-Nesi, and Wagner L. Araújo. 2020. "Engineering Improved Photosynthesis in the Era of Synthetic Biology." *Plant Communications* 1, no. 2 (March 9): 1–17. https://www.sciencedirect.com/science/article/pii/S2590346220300134#abs0010.

Bloom, Nicholas A., Charles I. Jones, John Van Reenen, and Michael Webb. 2020. "Are Ideas Getting Harder to Find?" *American Economic Review* 110, no. 4 (April): 1104–44.

Clancy, Matthew, Keith Fuglie, and Paul Heisey. 2016. "U.S. Agricultural R&D in an Era of Falling Public Funding." *Amber Waves*, November 10, 2016. US Department of Agriculture, Economic Research Service.

Clark, J. A., J. H. Martin, and C. R. Ball. 1922. *Classification of American Wheat Varieties*. US Department of Agrigulture Bulletin no. 1074, November 8, 1922. Revised August 1923. Washington, DC: Government Printing Office.

Clemens, Michael A., Ethan G. Lewis, and Hannah Postel. 2018. "Immigration Restrictions as Active Labor Market Policy: Evidence from the Mexican Bracero Exclusion." *American Economic Review* 108 (6): 1468–87.

Conley, Timothy G., and Chris Udry. 2010. "Learning about a New Technology: Pineapple in Ghana." *American Economic Review* 100, no. 1 (March): 35–69. https://www.aeaweb.org/articles?id=10.1257/aer.100.1.35.

Ehrenberg, R. 2012. "American Higher Education in Transition." *Journal of Economic Perspectives* 26:193–216.

Evenson, Robert E. 1978. "A Century of Productivity Change in U.S. Agriculture: An Analysis of the Role of Invention, Research and Extension." Center Discussion Paper No. 296. New Haven, CT: Yale University, Economic Growth Center.

———. 1989. "Spillover Benefits of Agricultural Research: Evidence from U.S. Experience." *American Journal of Agricultural Economics* 71 (2): 447–52.

Ewens, Michael, Ramana Nanda, and Matthew Rhodes-Kropf. 2018. "Cost of Experimentation and the Evolution of Venture Capital." NBER Working Paper No. 24523. Cambridge, MA: National Bureau of Economic Research.

Foster, Andrew D., and Mark R. Rosenzweig. 1995. "Learning by Doing and Learning from Others: Human Capital and Technical Change in Agriculture." *Journal of Political Economy* 103 (6): 1176–209.

Fuglie, Keith O., and Paul W. Heisey. 2007. *Economic Returns to Public Agricultural Research*. Economic Brief No. 6388, US Department of Agriculture, Economic Research Service.

Gallardo, R. Karina, and Johannes Sauer. 2018. "Adoption of Labor-Saving Technologies in Agriculture." *Annual Reviews of Resource Economics* 10:185–206.

Griliches, Zvi. 1957. "Hybrid Corn: An Exploration in the Economics of Technological Change." *Econometrica* 25 (4): 501–22. https://doi.org/10.2307/1905380.

Guttman, J. 1978. "Interest Groups and the Demand for Agricultural Research." *Journal of Political Economy* 86:467–84.

Hayami, Y., and V. W. Ruttan. 1970. "Factor Prices and Technical Change in Agricultural Development: The United States and Japan, 1880–1960." *Journal of Political Economy* 18:1115–41.

Hicks, J. 1932. *The Theory of Wages*. London: Macmillan.

Huffman, W. E., and R. E. Evenson. 2006. "Do Formula or Competitive Grant Funds Have Greater Impacts on State Agricultural Productivity?" *American Journal of Agricultural Economics* 88:783–98.

Hurley, Terrance M., Xudong Rao, and Philip Pardey. 2014. "Re-examining the Reported Rates of Return to Food and Agricultural Research and Development American." *Journal of Agricultural Economics* 96 (5): 1492–1504.

Kantor, Shawn, and Alexander Whalley. 2019. "Research Proximity and Productivity: Long Term Evidence from Agriculture." *Journal of Political Economy* 127 (2): 819–54.

Lobell, D. B., M. J. Roberts, W. Schlenker, N. Braun, B. B. Little, R. M. Rejesus, and G. L. Hammer. 2014. "Greater Sensitivity to Drought Accompanies Maize Yield Increase in the U.S. Midwest." *Science* 344 (6183): 515–19.

Lobell, David B., Wolfram Schlenker, and Justin Costa-Roberts. 2011. "Climate Trends and Global Crop Production since 1980." *Science*, 333 (6042): 616–20. https://doi.org/10.1126/science.1204531.

McFadden, Jonathan, David Smith, Seth Wechsler, and Steven Wallander. 2019. "Development, Adoption, and Management of Drought-Tolerant Corn in the United States." Economic Research Service, *Economic Information Bulletin* no. 204, January 2019.

Moser, Petra. 2012. "Innovation without Patents—Evidence from World's Fairs." *Journal of Law and Economics* 55 (1): 43–74.

Moser, Petra, and Paul W. Rhode. 2012. "Did Plant Patents Create the American Rose?" In *The Rate and Direction of Technological Change*, edited by Joshua Lerner and Scott Stern, 413–41. Chicago: University of Chicago Press.

Mowery, D. C., R. R. Nelson, B. N. Sampat, and A. A. Ziedonis. 2004. *Ivory Tower and Industrial Innovation: University-Industry Technology Transfer before and after the Bayh-Dole Act*. Palo Alto, CA: Stanford University Press.

Olmstead, Alan L., and Paul W. Rhode. 2002. "The Red Queen and the Hard Reds: Productivity Growth in American Wheat, 1800–1940." *Journal of Economic History* 62 (2): 929–66.

———. 2008. *Creating Abundance: Biological Innovation and American Agricultural Development.* Cambridge: Cambridge University Press.

Pardey, Philip G., and Nienke M. Beintema. 2001. *Slow Magic: Agricultural R&D a Century after Mendel.* Agricultural Science and Technology Indicators Initiative (ASTI). Washington, DC: International Food Policy Research Institute (IFPRI).

Pardey, Philip G., Nienke M. Beintema, Steven Dehmer, and Steven Wood. 2006. *Agricultural Research: A Growing Global Divide?* Agricultural Science and Technology Indicators Initiative (ASTI). Washington, DC: International Food Policy Research Institute (IFPRI).

Perkmann, Markus, Valentina Tartari, Maureen McKelvey, Erkko Autio, Anders Broström, Pablo D'Este, Riccardo Fini et al. 2013. "Academic Engagement and Commercialisation: A Review of the Literature on University–Industry Relations." *Research Policy* 42 (2): 423–42.

Roberts, Michael J., and Wolfram Schlenker. 2011. "The Evolution of Heat Tolerance of Corn: Implications for Climate Change." In *The Economics of Climate Change: Adaptations Past and Present*, edited by Gary D. Libecap and Richard H. Steckel, 225–51. Chicago: University of Chicago Press.

Rosenberg, Nathan, and Richard Nelson. 1994. "American Universities and Technical Advance in Industry." *Research Policy* 23 (3): 323–48. https://EconPapers.repec.org/RePEc:eee:respol:v:23:y:1994:i:3:p:323–48.

Sampat, B. 2006. "Patenting and US Academic Research in the 20th Century: The World before and after Bayh-Dole." *Research Policy* 35:772–89.

Sampat, Bhaven N., Frank R. Lichtenberg. 2011. "What Are the Respective Roles of the Public and Private Sectors in Pharmaceutical Innovation?" *Health Affairs* 30 (2): 332–39. https://doi.org/10.1377/hlthaff.2009.0917.

San, Shmuel. 2020. "Labor Supply and Directed Technical Change: Evidence from the Abrogation of the Bracero Program in 1964." Working paper, New York University, New York, NY, November 9, 2020. https://mulysan.github.io/San_bracero.pdf.

Schlenker, Wolfram, and Michael J. Roberts. 2009. "Nonlinear Temperature Effects Indicate Severe Damages to U.S. Crop Yields under Climate Change." *Proceedings of the National Academy of Sciences* 106 (37): 15594–98. https://doi.org/10.1073/pnas.0906865106.

Sengupta, Abhijit, and Amit S. Ray. 2017. "University Research and Knowledge Transfer: A Dynamic View of Ambidexterity in British Universities." *Research Policy* 46 (5): 881–97.

Shih, Tiffany M., and Brian D. Wright. 2011. "Agricultural Innovation." In *Accelerating Energy Innovation: Insights from Multiple Sectors*, edited by Rebecca M. Henderson and Richard G. Newell, 49–85. Chicago: University of Chicago Press.

Stackman, E. C., Richard Bradfield, and Paul Mangelsdorf. 1967. *Campaigns against Hunger*. Cambridge, MA: Belknap Press of Harvard University Press.

Suri, Tavneet. 2011. "Selection and Comparative Advantage in Technology Adoption." *Econometrica* 79 (1): 159–209.

Sutch, Richard. 2011. "The Impact of the 1936 Corn Belt Drought on American Farmers' Adoption of Hybrid Corn." In *The Economics of Climate Change: Adaptations Past and Present*, edited by Gary D. Libecap and Richard H. Steckel, 195–223. Chicago: University of Chicago Press.

Tartari, V., M. Perkmann, and A. Salter. 2014. "In Good Company: The Influence of Peers on Industry Engagement by Academic Scientists." *Research Policy* 43:1189–203.

Tartari, V., and A. Salter. 2015. "The Engagement Gap: Exploring Gender Differences in University—Industry Collaboration Activities." *Research Policy* 44: 1176–91.

Thursby, J., and M. Thursby. 2011. "Has the Bayh-Dole Act Compromised Basic Research?" *Research Policy* 40:1077–83.

Wright, Brian D. 2012. "Grand Missions of Agricultural Innovation." *Research Policy* 41 (10): 1716–28.

1

The Roots of Agricultural Innovation
Patent Evidence of Knowledge Spillovers

Matthew Clancy, Paul Heisey, Yongjie Ji, and GianCarlo Moschini

1.1 Introduction

Changes in the technology of farming have profoundly affected US production agriculture over the past century (Gardner 2002). Myriad innovations adopted by farmers contributed to this transformation, including mechanization; vastly improved genetics for plants and animals; novel inputs such as fertilizers, pesticides, and antibiotics; and the reorganization of farming activities to exploit specialization and scale economies. The results are impressive: between 1950 and 2015, for example, the total factor productivity index for US agriculture increased by 167 percent compared to 97 percent for the US nonfarm private sector.[1]

Digging deeper into the causes of these waves of agricultural technical change uncovers the critical role played by past research and development

Matthew Clancy is an assistant teaching professor in the Department of Economics at Iowa State University.

Paul Heisey was at the Economic Research Service, US Department of Agriculture, when this research was conducted.

Yongjie Ji is an assistant scientist at the Center for Agricultural and Rural Development at Iowa State University.

GianCarlo Moschini is a professor and the Pioneer Endowed Chair in Science and Technology Policy in the Department of Economics at Iowa State University.

An earlier version of this chapter was presented at the NBER conference on the "Economics of Research and Innovation in Agriculture," Washington, DC, May 17, 2019. For acknowledgments, sources of research support, and disclosure of the authors' material financial relationships, if any, please see https://www.nber.org/books-and-chapters/economics-research-and-innovation-agriculture/roots-agricultural-innovation-patent-evidence-knowledge-spillovers.

1. Agricultural total factor productivity data based on input, output, and productivity data published by the USDA's Economic Research Service (ERS; USDA ERS 2020). US nonfarm private sector total factor productivity taken from table XG4–2 from US Bureau of Labor Statistics (2007).

(R&D) activities. Griliches's (1957) pioneering work on the yield improvements due to hybrid maize found a large payoff to the cumulated past research investment in this technology: a benefit-cost ratio of 7, or an internal rate of return of about 40 percent. More broadly, for a set of studies published over the 1965–2005 period, the median estimate of the internal rate of return of agricultural R&D was 45 percent, or a benefit-cost ratio of about 10 (Fuglie and Heisey 2007).

R&D explicitly focused on agriculture, conducted by firms and public organizations, is obviously essential to agricultural innovation. Nonagricultural R&D, however, may also play a role via so-called knowledge spillovers. The most immediate output of R&D is new knowledge, but it has long been recognized that the R&D performed by one entity (e.g., a public lab, or a firm) in a given industry may have substantial productivity impacts outside this entity or industry (Griliches 1992). At a positive level, spillovers create serious challenges to the task of inferring, from data, what R&D effort had which effect on outcomes of interest.

In this chapter, we focus squarely on assessing the extent to which knowledge spillovers may impact agricultural innovation. With some caveats, discussed later, we find that our proxies for knowledge flows—citations to patents, citations to scientific papers, and novel text—suggest that more than 50 percent of knowledge spillovers originate in nonagricultural knowledge domains.

Knowledge spillovers have received limited attention in previous agricultural R&D studies. The typical econometric procedure has been to regress an estimate of agricultural productivity on relevant past R&D expenditures. To account for spillovers, some studies include broader measures of R&D expenditures. Attention has mostly concerned spillover between segments of agricultural R&D (Evenson 1989), or privileged spatial R&D spillovers—that is, across states or countries (Latimer and Paarlberg 1965; Khanna, Huffman, and Sandler 1994). Alston (2002) concludes that such spillovers are sizeable: interstate or international R&D spillovers may account for more than half of the measured agricultural productivity growth. Consideration of vertical spillover effects in agriculture is rare. One exception is Wang, Xia, and Buccola (2009), who relate public research in three life science fields (biology, agriculture, and medicine), and private research in two of these fields (agriculture and medicine), to research output (measured by patents) of private firms in agriculture and medicine.[2]

This chapter's contribution is to provide new methods and data on the scope of knowledge spillovers in agriculture. In contrast to most studies in this area, we do not attempt to calculate the rate of return on R&D. Instead,

2. Consistent with our results, Wang, Xia, and Buccola (2009) find evidence of substantial spillovers from upstream biological to downstream agricultural and medical science and from the public to the private sector in both downstream agriculture and downstream medicine.

we measure the extent of knowledge spillovers by directly observing proxies for knowledge flows and measuring the share of these that originate in nonagricultural knowledge domains. The goal is to provide new evidence on the extent to which agricultural technologies draw on knowledge originally developed outside of agriculture. We do so by developing various knowledge-flow proxies embedded in US agricultural patents granted over the 1976–2018 period.

Our initial step is to identify a set of relevant agricultural patents among the universe of US patents granted over this period. Note that while our analysis is restricted to US patents, these patents proxy global agricultural research. Depending on the subsector, we estimate that 14 percent to 49 percent of US agricultural patents are based on foreign research. We identify patents belonging to six distinct subsectors of agriculture: animal health, biocides, fertilizer, machinery, plants, and agricultural research tools (hereafter shortened to "research tools"). We chose these six subsectors because we can identify their patents with relative precision and because, while not exhaustive, they span the major biological, chemical, and mechanical technology fields that have contributed to productivity growth in agriculture. Given significant differences in the size, organization, and scientific-technological knowledge base of these subsectors, our results are consistently presented separately for the six subsectors of interest. We then track the knowledge roots of each patent, using three proxies embedded in the patent document.

The first proxy we consider is citations to prior patents, which provides a measure of how agricultural innovations build on other (patented) technologies. When the cited patent is not an agricultural patent, this provides direct evidence of a knowledge spillover from outside agriculture. Furthermore, one advantage of studying citations to patents is that we can also identify the assignee of the cited patent. A major part of our work is to determine the "agricultural focus" of assignees' R&D based on the share of agricultural patents in the assignee's recent patent portfolio. This permits us to go beyond the binary classification of whether a cited patent is agricultural and instead characterize it based on the agricultural focus of its assignee. For example, we can measure whether a cited patent belongs to a firm that generally specializes in agricultural R&D or belongs to an assignee that has zero agricultural patents.

The second proxy we employ is citations to scientific journal articles, which provides a measure of how agricultural innovations build on prior research. Citations to the scientific literature are important as a way of capturing the impact of public sector research, because public sector research frequently does not result in a patent. We create a classification system for scientific journals, identifying agricultural science, other biology, other chemistry, and "other" journals. We interpret a citation to a nonagricultural journal as evidence of a knowledge spillover to agriculture from outside of its natural knowledge domain.

Whereas citations to prior patents are generally acknowledged to contain both signal and noise, there is debate about the relative magnitude of each. For example, Chen (2017) finds the textual similarity of patents to their citations is much higher than to a control. An early survey by Jaffe, Trajtenberg, and Fogarty (2000), however, found that only 38 percent of respondents were aware of the cited patent before or during the invention. Other papers have also found evidence that citations may not reflect genuine knowledge flows (Lampe 2012; Moser, Ohmstedt, and Rhode 2018).[3] For this reason, we also develop a third method of measuring knowledge flows based on the text of patents.

The text analysis we develop identifies words and phrases that are new and important in agricultural patents applied for in the second half of our data sample (1996–2018). We call these words and phrases "text-novel concepts" and identify more than 100 in each subsector. We then scan the text of the entire patent corpus for prior patents (outside the agricultural subsector) that also mention these text-novel concepts. For example, in animal health, the word *pyrimethamine* and the phrase *equine protozoal myeloencephalitis* do not appear in any animal health patents prior to 1996 but are relatively common thereafter. In this case, we interpret prior mentions of *pyrimethamine* in human health patents prior to 1996 as evidence of a knowledge spillover from outside agriculture.

Our main finding is that knowledge spillovers from outside agriculture are important and influential for agricultural R&D, possibly as much as knowledge generated within agricultural science domains. In three of the subsectors studied—animal health, fertilizer, and machinery—every one of our proxies for knowledge flows originates outside of agriculture more than 50 percent of the time. In two additional sectors—biocides and research tools—we have mixed evidence, but the majority of our proxies still originate outside of agriculture over 50 percent of the time. Only in the plants subsector do we typically find most knowledge flows point to agricultural technologies and research, though even this is not unanimous.

A second finding is that the nonagricultural domains that are important sources of knowledge spillover to agriculture are, in some sense, "close" to agriculture. It is typically more common for agricultural patents to cite or share text-novel concepts with the (nonagricultural) patents of firms that have at least one agricultural patent, even though the majority of patents belong to assignees with zero agricultural patents. Likewise, it is more common for citations to nonagricultural science journals to go to biology and chemistry journals than other journals, even though other journals account for the majority of journals.

Lastly, we demonstrate how text analysis can be a useful complement to

3. In general, there seems to be less cause for concern about bias in citations to the scientific literature (Roach and Cohen 2013).

citation-based measures of knowledge flows. In some cases, our text analysis suggests areas where citation-based results may be misleading. For example, in the biocides sector, the majority of patent citations flow to nonagricultural patents and journals. However, we find the majority of text-novel concepts for this sector (typically chemical names) have no prior mention outside of agriculture. It seems many of these chemicals appear for the first time in the patent corpus as part of a biocide patent.

Our text analysis, in principle, has the ability to capture much deeper roots of knowledge than citations. It may be that an idea developed originally far outside of agriculture eventually enters agriculture via a long chain of citations. Because citation-based measures of knowledge spillovers only observe the last step in such chains, when an agricultural patent makes citations, these citations may understate the role of nonagricultural knowledge spillovers. In contrast, our text analysis lets us track *all* prior mentions of important concepts used in agricultural R&D, including mentions that are many steps removed from agriculture. Consistent with this notion, we find the share of text-novel concepts that originate in agricultural patents is smaller when measured using text rather than citations (except in the biocides sector).

The rest of the chapter is organized as follows: Section 1.2 describes our methodology for generating data on agricultural R&D output, knowledge flows, and originating knowledge domain and gives an example. Section 1.3 presents our main results. Section 1.4 discusses these results, and section 1.5 establishes that they are robust to a series of alternative assumptions. Section 1.6 concludes with some directions for future research.

1.2 Data

Our goal is to measure the extent of knowledge spill-ins for agricultural R&D. To accomplish this, we require three elements: a measure of agricultural research output, a measure of knowledge flows, and a measure of originating knowledge domain. These three components, plus our proxies for them, are illustrated in figure 1.1.

Working from right to left, to measure agricultural research output, we use patents with primarily agricultural applications. Our chapter focuses on six agricultural subsectors: animal health, biocides, fertilizer, machinery,

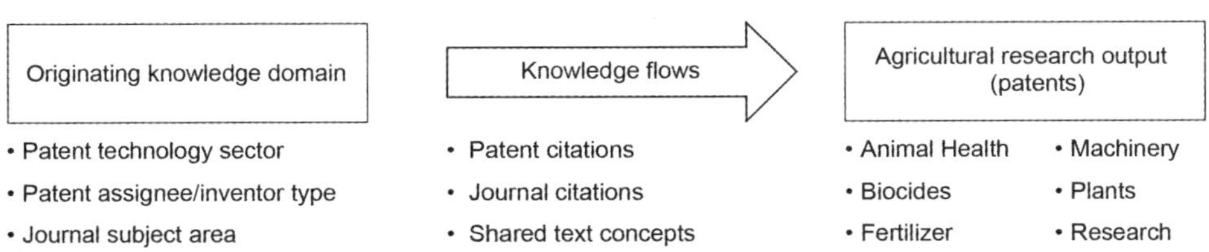

Fig. 1.1 Knowledge spill-ins and proxy elements

plants, and research tools. We describe our method for identifying these patents in section 1.2.1. We measure knowledge flows in three ways: patent citations to other patents, patent citations to academic journals, and shared patent text. We describe how we generate these three proxies in section 1.2.2. We also define the originating knowledge domain in three ways: with patent technology classes, with assignee type, and with journal subject areas. We describe these methods in section 1.2.3. Section 1.2.4 provides some brief summary statistics for our data.

1.2.1 Measuring Agricultural Research Output

We use the universe of US patents granted between 1976 and 2018 for our analysis, though for some subsectors, we only have data through 2015. Over this period, 5,886,981 patents were granted. While we use this entire data set in our analysis, we are particularly interested in the subset of patents closely related to agriculture. Conceptually, our guiding principle is to identify patents over technologies used primarily in either agricultural production or agricultural research. We attempt to exclude patented technologies that have many applications but where agriculture is not the primary use. For example, the CRISPR gene-editing technology has applications in agriculture but also many more applications in human medicine and fundamental research. We include only the subset of CRISPR patents that is closely related to agricultural research.

Our analysis is focused on six agricultural subsectors where we are able to identify related patents with relatively high precision: animal health, biocides, fertilizer, machinery, plants, and research tools. While we feel that these capture a large share of the major technological developments in agriculture over the last 40 years, we do not claim our analysis is exhaustive. In particular, the livestock genetics sector does not rely on patent protection to the same extent that the crop genetics sector does, and so we lack any information on this important sector. Another notable sector we are missing is information technology (e.g., software) applied to agriculture, for which we lack reliable means of distinguishing software with primarily agricultural application from others. Also note that our analysis does not extend to the processing of agricultural products, either into food, feed, or biofuel.

With one exception (described below), our classification of patents starts with the cooperative patent classification (CPC) system. The CPC system is used by the US Patent and Trademark Office (USPTO) to classify patents into different technology categories in order to facilitate USPTO patent examiners (and other interested parties) in finding relevant prior art. We use the *cpc_current* file, available on the USPTO's PatentsView website, as our primary source. Patents are generally assigned multiple classifications, but we use only the primary classification for the purpose of allocating patents to a particular group.

For the biocide, fertilizer, and machinery subsectors, we identify CPC codes associated with the relevant sector and assign patents with identified codes as their primary classification to the relevant sector. A complete list of patents by subsector is available in the supplemental materials. Here we briefly describe our approach:

Biocides: This subsector includes fungicides, herbicides, insecticides, pesticides, and other chemicals meant to control biological pests. We start with CPC classification A01N, which includes these chemicals as well as chemicals for the preservation of bodies. We include any classifications under A01N related to biocides but exclude classifications related to the preservation of bodies (which tend to begin with A01N 1/).

Fertilizer: This subsector includes chemical fertilizers. We use CPC classifications beginning with C05, which corresponds to chemical fertilizer technology.

Machinery: This subsector includes agricultural machinery, with a focus on mechanically powered machinery. Within the CPC classification A01, we include any classification related to agricultural machinery (e.g., harvesting, mowing, planting, milking) and exclude many other categories unrelated to machinery (e.g., structures, forestry, fishing, hunting, and most of the other agricultural subsectors considered). Most of our agricultural machinery patents are classified under A01B, A01C, A01D, and A01F. Within the machinery categories, we also exclude classifications related to hand tools and animal-driven machinery.

These three subsectors require no additional processing. For the plant cultivar and agricultural research tools subsectors, the CPC classification system is not sufficiently precise for our purposes, so we supplement the CPC approach with manual cleaning.

Plants: This subsector includes utility patents for specific plant varieties/cultivars.[4] We begin with the set of patents assigned primary CPC code A01H, which includes both patented plant cultivars and plant modification and reproduction techniques as well as related technologies. We exclude CPC codes related to nonagricultural plants and fungi. From the remaining set, we manually identify patents for plant cultivars by inspecting the patent title, abstract, and claims.

Biological research tools: This subsector (hereafter shortened to "research tools") includes technologies for conducting biological research—for example, genetic engineering and traditional breeding techniques. We begin with CPC classifications under the category A01H that are related to processes for modifying agricultural plants and add some classifications under CPC class C12N (microorganisms and enzymes) that are specifically designated as being for the modification of plants. Note that

4. Note that this subsector does not include "plant patents," a distinct form of intellectual property dating to 1930 and applicable to asexually reproduced plants (Clancy and Moschini 2017).

A01H also includes plant cultivar patents; we exclude any patents that are already classified in the plants subsector.

Animal health: This subsector includes all patents associated with medical technologies approved for use in veterinary medicine by the US Food and Drug Administration (FDA).

To obtain data on animal health patents, we adopt a different approach than that used for the other subsectors. While the CPC system suffices to identify patents related to medical technology, it does not distinguish between medical technologies for human application and those for non-human animal application. Instead, to identify patents for veterinary medicine technologies, we rely on FDA archival data. To facilitate generic competition in the animal health market, since 1989 the FDA has maintained a list of patents associated with all approved veterinary medicine products. Using archival records of this list, Clancy and Sneeringer (2018) develop a list of all patents associated with approved veterinary medicine products.

It should be noted that the patents in the animal health subsector are subject to a selection effect that is not present in the other sectors. This is because animal health patents are only included if they are associated with veterinary drugs that eventually receive FDA approval. Drugs that are not approved may have associated patents, and we miss these. This selection effect may bias our results for this subsector in two ways. First, if successful and unsuccessful drugs enjoy spill-ins at differential rates, our results will only apply to successful drugs. In our robustness checks, however, we find little evidence in other subsectors that the most valuable patents differ dramatically in their citation patterns. Second, and perhaps more importantly, by omitting patents associated with unsuccessful drug applications, we will misclassify citations to these patents as citations to nonagricultural patents. This may partially account for our finding that animal health relies more on nonagricultural knowledge flows than other agricultural subsectors (although there are, of course, other plausible explanations for such a finding).

Figure 1.2 illustrates the annual number of (granted) patents, by application year, in each of these subsectors. A few preliminary observations are in order. First, most subsectors exhibit a sharp decline in patents in the last few years of the sample. This is due to a truncation effect: we only observe patents if they are granted by 2016 in most sectors (we have data until 2018 for our plants and research tools subsectors), and few patents applied for in 2014 and 2015 are granted by 2016.

Second, the plants and research tools subsectors exhibit a sharp increase from zero (or close to zero) in the 1980s. This is due to legal changes in the patentability of biological innovation in the wake of the 1980 *Diamond v. Chakraborty* Supreme Court case (Clancy and Moschini 2017). Prior to 1980, biological innovations such as new plant varieties were not patentable subject matter. It is important to note that any R&D related to biological

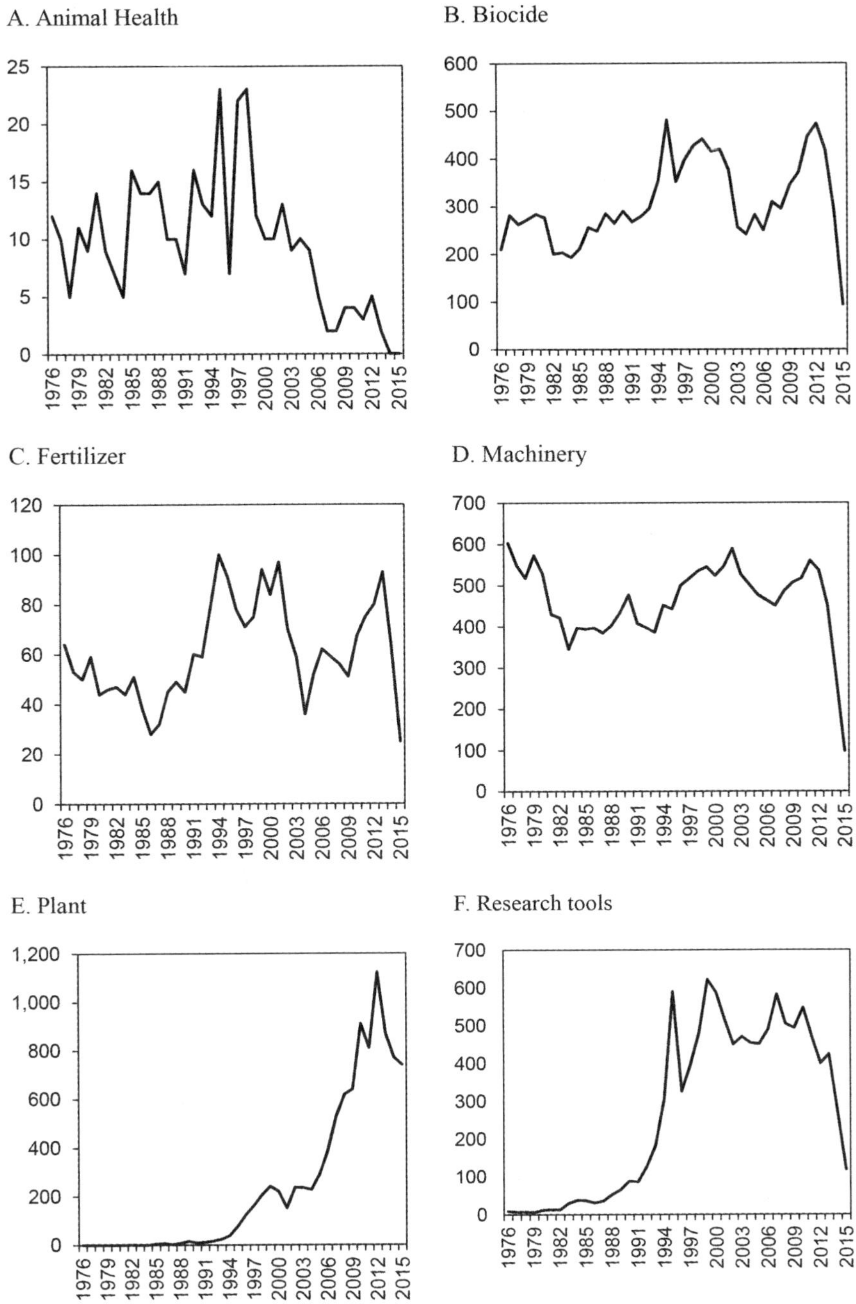

Fig. 1.2 Number of granted patents by application year and subsector, 1979–2016

innovation that occurs prior to 1980 is unlikely to be reflected in the patent record.

Finally, note that the scale of the vertical axis in figure 1.2 varies substantially across sectors. In our data set, the animal health sector has the smallest number of patents (414), and the machinery subsector has the most (19,362). Because of the variability in the size of the subsector, how long innovation in the subsector has been eligible for patent protection, and the presence of selection effects in the animal health subsector, in this chapter, we always report disaggregated results by subsector.

1.2.2 Measuring Knowledge Flows

Our first measure of knowledge flows is patent citations to other patents. We use the USPTO PatentsView data set *uspatentcitation* as our source for patent citations. This provides the patent number of both the citing and cited patent, and identifies who added the citation (the applicant, examiner, or other parties), from 2002 onward. Because we will be aggregating cited patents into different sectors and assignee types, we limit ourselves to citations to patents granted between 1976 and 2016.

Our second measure of knowledge flows is patent citations to academic journals. We estimate public sector patents are just 2 percent of all patents granted in our observation period, far below the public sector's share of R&D (agricultural or otherwise). Accordingly, to measure the role of public sector R&D, it is important to supplement our patent citation analysis with journal citations. Analysis of citations to nonpatent literature is complicated by the absence of standardized citation formatting. Patent applicants cite articles in a wide variety of ways: with or without abbreviations, with commas or periods to divide information, and with differences in the order of author names, year, title, journal, volume number, and so on. An emerging literature is attempting to match the raw citation text in patent documents to standardized journal entries in databases such as Clarivate (formerly Thompson Reuters) Web of Science, Elsevier Scopus, Google Scholar, Crossref, PubMed, and the Microsoft Academic Graph (MSAG). We use Marx and Fuegi (2019), a data set based on text-analysis algorithms that matches raw patent text to entries in the MSAG. Marx and Fuegi (2019) estimate they capture 90 percent of citations with 99 percent accuracy.

Our third measure of knowledge flows is a novel use of patent text, extending approaches pioneered by Packalen and Bhattacharya (2015) and Balsmeier et al. (2018). We identify a large set of text-novel concepts—proxied by one-, two-, and three-word strings of text—that are popular in agricultural innovation in the second half of our data set but absent from the first half. We find all mentions of these text-novel concepts in other patents and use earlier mentions of the concept as a measure of potential knowledge flow. Because this approach is novel, we describe it in some detail here.

The goal of this approach is to identify strings of text in patents that proxy for concrete ideas and concepts with technological applications. Following Packalen and Bhattacharya (2015), we define a "concept" as a text string consisting of one, two, or three words, without separating punctuation between them (i.e., hyphens are permitted).

For a given agricultural subsector's patents, we break the text of the title, abstract, and claims into concepts. This includes all individual words as well as all sequences of two or three words, as long as the words are not divided by punctuation (with the exception of hyphens). We focus on the title, abstract, and claims because these likely are most informative as to the important concepts in a patent: titles and abstracts are meant to succinctly describe the innovation, while claims are legally binding.

We next clean the text of these concepts, using an approach similar to Packalen and Bhattacharya (2015). We convert all text into lowercase letters. We then exclude concepts with numbers as one of the words, concepts where words are divided by punctuation, or concepts that are unusually short and long (in terms of their total number of characters).[5]

This leaves us with a very large set of text, most of which does not correspond to ideas and concepts with technological applications. To focus on new ideas in agriculture, we next divide our data set in half. The concepts in patents applied for in the first half of our observation period (1976–96) form a *baseline dictionary*. The concepts in patents applied for in the second half of our observation period (1996–2016) form a set of *recent concepts*. Any recent concept that is not contained in the baseline dictionary is considered a novel concept. Intuitively, this is a string of text that did not appear in any of the subsector's patent abstracts, titles, or claims prior to 1996 but does appear after 1996.

Next we calculate the number of subsector patents that contain each novel concept in the abstract, title, or claim. We call these "mentions." For example, the word *trimethoprim* refers to an antibiotic. It does not appear in any animal health patents prior to 1996 but appears in eight patents after 1996. We therefore say *trimethoprim* is a novel concept with eight mentions.

Our goal is to identify a set of important agricultural concepts. To do this, we first identify the 200-plus novel concepts with the most mentions. We frequently identify more than 200 concepts in this first pass, because mentions are necessarily integers and usually there are multiple concepts with the same number of mentions as the 200th concept. By construction, these are strings of text that did not appear in any of the sector's patent abstracts, titles, or claims prior to 1996 but were relatively common after 1996.

To increase our confidence that our concepts are good proxies for concrete ideas and concepts with technological applications, we go beyond Packalen

5. Following Packalen and Bhattacharya (2015), we exclude one-word concepts shorter than 3 characters or longer than 29 characters, two-word concepts shorter than 7 characters or longer than 59 characters, and three-word concepts shorter than 11 characters or longer than 89 characters. We also exclude concepts that include words in the Python NLTK stop words list.

and Bhattacharya (2015) and Balsmeier et al. (2018) and manually clean the set of candidate concepts using these four guidelines. We exclude the following:

1. Concepts with numbers and measurements: These are unlikely to correspond to generalizable ideas or concepts, as they usually refer to specific measurements that are not good proxies in the absence of more context (e.g., "90 degrees," "1,500 mL").
2. Connective phrases: These are largely free of concepts and ideas with technological applications and instead likely reflect variation in preferred patent language (e.g., "combinations thereof," "one particular type").
3. Words with multiple context-dependent meanings: When a set of words can have significantly different meanings in different contexts, then it is a poor proxy for our purposes because it may be mentioned in multiple patents with no technological similarity (e.g., "artificial," which could be paired with "intelligence," "insemination," or "sunlight").
4. Concepts including uninformative words: If some of the words in a concept appear to be valid (not excludable by any other criteria), but they only appear in conjunction with an additional word that is uninformative (e.g., "said" or "and"), we exclude the concept. In these cases, it is likely that the concept is not really novel—only the conjunction of the concept and the uninformative word (e.g., "said data structure," "the database" [if "data structure" and "database" do not appear as novel concepts themselves, then they were in use in 1976–96; only the exact formulation adding "said" or "the" was not]).

Three of the coauthors independently examined the list of candidate concepts using the foregoing four criteria, and any concept excluded by at least two of the three coauthors was removed. These exclusion criteria remove 37 percent of the top 200 concepts overall, with a low of 11 percent in biocides and a high of 47 percent in machinery. As a robustness check, we reperform our analysis on the set of concepts that are retained unanimously by all these coauthors. What remains constitutes our set of text-novel concepts. They form a set of text proxies for concrete technological ideas that are important in agricultural innovation over the 1996–2016 period and are new at least in the sense that they were not used over 1976–96 in patents. In some cases, the underlying concepts are not actually new but represent one of two things: first, the discovery of new applications for ideas that had been in a state of dormancy over 1976–96; and second, an expansion of the use of technological terms from the scientific literature to patent text. This latter phenomenon is often the result of an expansion of patentability, as in the case of utility patents for plant cultivars. For patents granted after 1996, depending on the subsector, anywhere from 17 percent (in machinery) to 94 percent (in plants) of patents mention one of the associated text-novel concepts. See table 1.5 for the breakdown by subsector.

Table 1.1 **Top 10 text-novel concepts by patent subsector, 1996–2016**

	Top 10 text-novel concepts
Animal health	Protozoal, trimethoprim, microbial, microbial infection, ear, preservative, terbinafine, penetration enhancer, kinase, bird
Biocides	Thiamethoxam, azoxystrobin, clothianidin, trifloxystrobin, spinosad, acetamiprid, thiacloprid, prothioconazole, pyraclostrobin, emamectin
Fertilizer	Selenium, itaconic, tea, canola, mean particle, chlorine dioxide, wetting agents, phosphite, ferrate, compost tea
Machinery	Controller configured, actuator configured, apparatus configured, antenna, dairy livestock, arm configured, flexible cutterbar assembly, controller operable, opening configured, gps receiver
Plants	Insect resistance, transgene, conversion, locus, trait selected, locus conversion, carbohydrate, backcross, metabolism, carbohydrate metabolism
Research tools	Clustal, one regulatory sequence, silencing, polynucleotide selected, isolated polynucleotides, chimeric gene results, polynucleotide operably linked, polynucleotide operably, polyunsaturated fatty acids,Rnai

The top 10 text-novel concepts in each subsector are listed in table 1.1. See the appendix for a complete list of top text-novel concepts in each subsector, broken down by those unanimously retained (the majority) and those retained only by two out of three reviewers (which are excluded in a robustness check). A cursory look at table 1.1 illustrates how text concepts align with our intuitions about the knowledge base in different fields: animal health, plants, and research tools all involve biological terms; biocides is mostly chemical names; machinery includes different mechanical components; and so on. In our main specification, we give equal weight to all concepts, but in our robustness checks, we show our results are robust to the clustering of concepts into families of related concepts.

To pinpoint potential knowledge flows, we identify any patents (whether agricultural or not) that mention these concepts. To do this, we again break the text of each patent's title, abstract, and claims into concepts; clean the text of these concepts; and identify any concepts that match the set of text-novel concepts in agriculture. These form the set of all patents (agricultural and otherwise) that mention any text-novel concepts in agriculture. We interpret such mentions as informative (albeit noisily) of knowledge flows and indicative that relevant research was ongoing in the sector to which agricultural researchers may have been exposed.

1.2.3 Originating Knowledge Domains

To measure the source of knowledge flows, we define the originating knowledge domain in three ways. Our first approach is simply to leverage our work of identifying patents in distinct agricultural subsectors. When a cited patent, or a patent linked by common text, belongs to one of our agricultural subsectors, we use the subsector as the originating knowledge domain. We find it useful, in general, to group these sectors by "own subsector" (e.g., an animal health patent citing another patent belonging to animal health), "other agriculture" (e.g., an animal health patent citing an agricultural research tools patent), and "not agriculture" (e.g., an animal health patent citing a human health patent).

1.2.3.1 Assignees

Our second approach relies on the assignees and inventors associated with patents. Most patents have an assignee, usually corresponding to the employer of one of the patent's inventors, and all patents have an inventor (or inventors). We are interested in distinguishing among assignees that are specialized in agriculture, assignees that conduct agricultural R&D but for whom it is not their primary focus, and assignees that conduct no agricultural R&D.

The problem of *assignee disambiguation* and *inventor disambiguation* in patents is an active area of research. In brief, this is the challenge of determining when two patents belong to the same assignee or inventor. What makes this challenging is that the USPTO does not assign unique IDs to inventors and assignees. Instead, assignees and inventors are listed as text in the patent document. The same set of text (e.g., "John Smith") may refer to different individuals/assignees. Or different text (e.g., "IBM" and "International Business Machines") may refer to the same individual/assignee.

We primarily rely on the disambiguation data set built by Balsmeier et al. (2018). These authors begin with the hand-curated NBER patent data project, which matched patents granted between 1976 and 2006 with publicly traded companies in the Compustat data set. Balsmeier et al. (2018) then use a k-nearest neighbor clustering algorithm for the remaining patents. This algorithm identifies the five assignees "closest" to the unmatched patent's assignee—in terms of having similar inventors, CPC codes, locations, and cited patents. It compares the assignee name of the unmatched patent to the names of these five nearest assignees and takes the closest match, provided the similarity of this match exceeds a threshold. Otherwise, a new assignee is added to the data set. A similar technique is used to disambiguate inventors.

We use Balsmeier et al. (2018) to differentiate between patents with assignees and those with individual inventors. However, assignees can take many forms: private firms, government agencies, nonprofit organizations, and even individuals different from the inventor who are assigned the patent. Balsmeier et al. (2018) do not distinguish between different kinds of assignees. We attempt to separate public sector assignees from private sector ones and then to characterize the extent of agricultural specialization for private sector assignees.

We adopt two approaches to identifying public sector assignees. First, the

USPTO's PatentsView *assignee* and *patent_assignee* files indicate whether an assignee is a government agency (state, federal or foreign). We classify the assignees of any patent with all government agency assignees as public sector assignees. Second, we use a list of keywords to identify major nongovernmental agency public sector assignees.[6] Any assignee that includes one of these keywords is also classified as a public sector patent.

Patents not classified as belonging to the public sector or individual inventors belong mostly to private sector firms. We are interested in characterizing the extent to which these firms' R&D focus is agricultural. We face two challenges here: ascertaining the extent of agricultural R&D and determining how to classify assignees that change their research focus over time. Some major firms dramatically reinvented themselves as agricultural companies over our observation period (Monsanto is a notable example), and so we need a way to distinguish between different phases of the firm's existence.

We use the share of patents classified as belonging to one of our agricultural subsectors to determine an assignee's agricultural focus. To capture the fact that assignees may change their research focus over time, we use only patents granted in the preceding five years to construct a time-varying, assignee-specific agricultural focus.[7] While we use this continuous measure of agricultural R&D focus, we also construct three types of assignee, where types can change from year to year:

Specialized agricultural assignee: A firm for which 50 percent or more of their patents, granted in the last five years, belong to one of our six agricultural subsectors.

Minority agricultural assignee: A firm that has at least one agricultural patent in the last five years but for which less than 50 percent of their patents, granted in the last five years, belong to one of our six agricultural subsectors.

Nonagricultural assignee: A firm with no patents granted in the last five years that belong to one of our six agricultural subsectors.

Our choice of five years balances two competing desires. A shorter time window introduces more noise into our estimates. A longer time frame is slow to recognize when a firm reorients its R&D focus. To assign firms a

6. Keywords include *university*, *universities*, *college*, *colleges*, institute of technology, foundation, school, *polytechnic institute*, *virginia tech*, *Argonne*, *Tulane education*, *board of regents*, *universita*, *universitat*, *universite*, and *universidad*.

We find these keywords largely match the number of patents granted to US colleges and universities, as reported by the USPTO and the NSF, in 2011 (USPTO Patent Technology Monitoring Team 2019).

7. When we do not have data on five prior years of patenting (i.e., in the first four years after an assignee begins to patent or the first four years in our data set), we use the patents granted in the first available five years or the maximum number of years available if five are not available. For example, for a patent granted in 1977, we use patents granted in 1976–80 to determine the assignee type in 1977.

position in the technology space, it is common to use the entire period under observation (see, e.g., Greenstone, Hornbeck, and Moretti 2010; Bloom, Schankerman, and Van Reenen 2013), and so our time five-year lag is relatively short. We find that using a longer time window results in fewer firms that we classify as specialized agricultural firms. Therefore, if we used a longer time frame, it would likely strengthen our conclusion that nonagricultural firms are a major source of knowledge flows in agriculture.

Approximately 5 percent of patents lack disambiguated assignee data in Balsmeier et al. (2018), and we assign these to an "unclassifiable" category. When a patent has multiple assignees spanning different types, we fractionally allocate the patent across different assignee types. Lastly, note that there is no concordance between assignees in the USPTO PatentsView data and the Balsmeier et al. (2018) data set. In the rare case (less than 1.5 percent) where a patent has multiple assignees, and some but not all are indicated as government agencies by the USPTO data sets, we cannot determine which of the assignees in Balsmeier et al. (2018) are the government agencies (text similarity matching fails). We allocate this small number of patents to the unclassifiable category.

Based on these criteria, 55 percent of all patents over our observation period belong to nonagricultural assignees, 23 percent belong to minority agricultural assignees, 15 percent belong to individuals, 5 percent are unclassifiable, 2 percent belong to public sector firms, and 0.5 percent belong to specialized agricultural firms. For comparison, patents in any of our agricultural subsectors account for 1 percent of all patents granted over the period. Note that this implies the agricultural patents of minority agricultural firms account for slightly more than 3 percent of their patents.

Table 1.2 displays the four assignees with the most patents in each agricultural subsector. As expected, they largely correspond to well-known firms.

1.2.3.2 Journal Classification

Our first two approaches to defining the originating knowledge domain are only appropriate for knowledge flows that are proxied by patents (i.e., either cited patents or patents with shared text concepts). Here we develop a third approach—appropriate for our journal citation proxy of knowledge flows—based on the classification of cited journals into broad academic categories. We create four main categories: agricultural science journals, other biology/biochemistry journals, other chemistry journals, and other journals.

Our list is based on the SCImago portal for the Scopus abstract and citation database for peer-reviewed literature.[8] Journals are placed in broad "subject areas," and within each subject area are more narrowly defined

8. See the website at https://www.scimagojr.com/.

Table 1.2 **Top four patent-holding assignees by subsector, 1976–2016**

	Top four assignees by patent holdings
Animal health	Pfizer Inc., Eli Lilly and Company, Alza Corporation, Hoechst Aktiengesellschaft
Biocides	Hoechst Aktiengesellschaft, BASF Aktiengesellschaft, Sumitomo Chemical Company Limited, CIBA Geigy Corporation
Fertilizer	Union Oil Company of California, Tennessee Valley Authority, OMS Investments Inc., Allied Signal Inc.
Machinery	Deere and Company, CNH America LLC, Unisys Corporation, J I Case Company
Plants	Pioneer Hi Bred International Inc., Monsanto Technology LLC, Stine Seed Farm Inc., Syngenta Participation AG
Research tools	Pioneer Hi Bred International Inc., E I Du Pont De Nemours and Company, Monsanto Technology LLC, The Regents of the University of California

Table 1.3 **Defining the set of agricultural sciences journals**

Agricultural and biological sciences	
Agricultural and biological sciences (misc.)	Journals manually inspected
Agronomy and crop science	All journals included
Animal science and zoology	Journals manually inspected
Aquatic science	Journals not inspected
Ecology, evolution, behavior, and systematics	Journals not inspected
Food science	Journals not inspected
Forestry	Journals not inspected
Horticulture	All journals included
Insect science	Journals manually inspected
Plant science	Journals manually inspected
Soil science	All journals included
Veterinary science	
Equine	Journals not inspected
Food animals	All journals included
Small animals	Journals not inspected
Veterinary (misc.)	Journals manually inspected

"subject categories." Journals can be placed in more than one subject category and, for that matter, in more than one subject area. To create the "agricultural science" category, we start with two SCImago subject areas: (1) "agricultural and biological sciences" and (2) "veterinary sciences." Table 1.3 lists the subject categories within these two areas and shows how the journals of each subject category are treated.

Note that because journals can be cross listed in several categories, it is possible for a journal to be designated as an agricultural science journal even if it belongs to one of the subject categories whose journals we do not inspect. This can occur, for example, if the journal is also listed in a category we do inspect. Eliminating duplicate entries results in a set of 981 journals classified as "agricultural sciences."

To create our set of "other biology/biochemistry" journals, we begin with all journals in the SCImago Agricultural and Biological Sciences area and Veterinary Sciences area that ended up not being included in the aforementioned agricultural sciences category. To this, we add all journals classified by SCImago in the "biochemistry, genetics, and molecular biology" subject area that were not already classified as Agricultural Sciences by us. This results in a set of 3,029 journals classified as "all other biology/biochemistry."

To create the "other chemistry" journal list, we combine all journals (not already classified in the preceding steps) from the "chemistry" and "chemical engineering" subject areas in the SCImago set. This results in a set of 995 journals classified as "other chemistry."

Lastly, all remaining journals in SCImago are classified as "other." It is important to note that this category contains several prominent multidisciplinary journals, such as *Science*, *Nature*, and *PNAS*. We show our results are robust to separating out these three major journals into their own category. In all cases, we retain journals, book series, and trade journals but mostly exclude conferences and proceedings volumes. This results in a set of 21,166 other journals.

A final challenge remains. Our source for journal citations is Marx and Fuegi (2019), which links the raw text in patents to entries in the MSAG. We match journal titles in the MSAG to journal titles in our SCImago classification system by a Levenshtein distance text-matching algorithm (we retain matches above 90 percent confidence). For "agricultural sciences," we further manually check all journal matches. Table 1.4 illustrates the share of MSAG journals that we successfully match to journals in the SCImago.

As indicated by table 1.4, we always match the majority of journals and typically match approximately 75 percent. Our performance is worse in the machinery subsector (60.9 percent)—this is probably due to the fact that this is a field where citations to academic journals are rare and citations to conference proceeding papers (which we mostly exclude) are common. In the plants subsector, the MSAG is unable to match 25 percent of nonpatent citations to journals. The manual inspection of a sample of these citations indicates that they mostly accrue to books, which are also not in our data set.

1.2.4 Summary

Table 1.5 provides a summary of our data.

Note that the subsectors vary significantly in their propensity to cite, especially with respect to nonpatent references (the majority of which are to academic journals). The machinery and fertilizer subsectors, for example,

Table 1.4 **Journal match performance**

	Matched to SCImago journals (%)	Matched in MSAG to other journals (%)	Not matched in MSAG to journals (%)
Animal health	75.6	16.9	7.5
Biocides	79.6	10.2	10.2
Fertilizer	74.1	11.9	14.0
Machinery	60.9	10.1	29.0
Plants	73.0	1.6	25.4
Research tools	92.4	3.5	4.1

Note: MSAG denotes Microsoft Academic Graph. Column (1) is the share of patent citations to journals in the MSAG that we match to journals in SCImago. Column (2) is the share of citations in the MSAG that Microsoft indicates correspond to journals but for which we are unable to match the entry to a journal in SCImago. Column (3) is the set of citations that Microsoft lacks enough information about to match to a journal.

Table 1.5 **Summary statistics**

	Patents	Share top four assignees (%)	Avg. patent cites made	Avg. nonpatent cites made	Share patents w/ text concepts (%)
Animal health	414	24.9	9.4	8.5	76.3
Biocide	12,774	13.7	8.3	6.5	24.2
Fertilizer	2,554	3.7	10.7	3.4	32.9
Machinery	19,362	16.8	13.2	1	16.7
Plants	10,216	67.0	7.6	9.2	94.4
Research tools	10,872	21.5	7.5	37.3	41.6

Note: "Patents" is the number of patents in the subsector. "Share top four assignees" is the share of these patents assigned to the four largest assignees. "Avg. patent cites made" is the mean number of citations made to other patents per patent. "Avg. nonpatent cites made" is the mean number of nonpatent references per patent. "Share patents w/ text concepts" is the share of patents granted after 1996 that mention one of the top text concepts included in our text analysis.

cite more patents than any other subsector but the fewest nonpatent references. Meanwhile, the research tools subsector cites nonpatent literature at more than four times the rate of the next highest subsector.

Subsectors also vary in their concentration. Whereas fertilizer patents are dispersed among a plethora of small assignees, plant patents are highly concentrated in a small number of firms (with Monsanto and Pioneer alone accounting for more than half of all patents). Table 1.5 also highlights how our text-analysis approach varies in how representative it is for different subsectors. Whereas the majority of patents granted after 1996 in animal health and plants carry one of our text-novel concepts, only 17 percent of such patents in machinery do (although, as the largest single subsector, the small share translates into thousands of patents).

While not a major focus of our chapter, the extent of foreign research in our data is also of interest and can be proxied by the share of foreign inventors on a given patent. Using the USPTO PatentsView inventors location data set, we classify patents as being derived from foreign research if all the inventors have non-US addresses and as being derived from domestic research if all the inventors have US addresses. For patents with some, but not all, inventors residing abroad, we classify the patent as fractionally foreign based on the share of its inventors that reside abroad (i.e., a patent with one out of two inventors residing abroad is listed as 0.5 foreign patents). By this measure, plants have the fewest foreign patents (14 percent) and biocides the most (49 percent). Table 1.A1 provides a full breakdown by subsector.

1.2.5 An Example

As an example, consider patent 5,747,476, titled "Treatment of Equine Protozoal Myeloencephalitis." The patent was applied for in July 1996, granted in May 1998, and assigned to the Mortar & Pestle Veterinary Pharmacy Inc. in Des Moines, Iowa. We classify this as an animal health patent. As the title suggests, it describes a novel treatment for equine protozoal myeloencephalitis (EPM), a debilitating neurologic disease that affects horses. At the time of the patent application, EPM was commonly treated by crushing two different kinds of tablets intended to treat humans—one with the active ingredient pyrimethamine and another with a trimethoprim-sulfonamide combination—and suspending the mixture in solution. This was given to the horse prior to feeding, often for 90 days. The patent describes a new therapy, designed specifically for EPM, that involves a compound of pyrimethamine and a sulfonamide ("preferably sulfadiazine") but with a much smaller dose of trimethoprim (or none at all).

Such an innovation obviously builds on ideas developed outside of agriculture. Pyrimethamine was discovered in 1952 and developed into an antimalarial treatment (for humans) in 1953 but has many applications in treating parasitic diseases. Sulfanomides have an even older history, forming part of the first set of antibiotics widely used (again, for humans) in the 1930s. However, their joint application in treating EPM is novel.

Patent 5,747,476 reflects these deep roots in several ways. It cites 10 patents, most of which have little to do with veterinary medicine (the oldest being US patent 4,293,547: "Method of Treating Malaria," granted in 1981). We identify patents as pertaining to agriculture if they belong in one of our agricultural patent data sets (which the cited ones do not) or if the assignees of these cited patents have other agricultural patents within the last five years. Where they do, we find that the share of agricultural patents is quite small. To take one example, patent 4,293,547 belongs to the Upjohn Company, and only 1.1 percent of its patents were agricultural in 1981 (over the preceding five years).

Only one cited patent belongs to a publicly owned entity: patent 5,486,535,

"Method of Treating Toxoplasmosis," which is assigned to the regents of the University of California. To understand the patent's use of publicly funded knowledge, we instead turn to its 13 citations to journals. The cited references include the *American Journal of Veterinary Research*, the *Canadian Veterinary Journal*, and the *Journal of Parasitology*. Of these, we classify the first two as agricultural science journals and the last as a biology/biochemistry journal, suggesting this patent draws on both specific agricultural research and basic biology.

Finally, the text of the patent itself contains important concepts. The word *pyrimethamine* is absent in our animal health data set for the first half of our observation period but relatively common in the second half, so it is one of our top text-novel concepts. The words *equine protozoal myeloencephalitis* represent another concept that is absent over 1976–95 but relatively common in animal health patents after 1995.

When we search the broader patent corpus for patents including the word *pyrimethamine* (in the title, abstract, or claims), we find many examples that predate its use in animal health (hardly surprising, given its history) not among the patents cited. These patents provide a third indicator that this patent draws on knowledge developed outside of agriculture. In contrast, the phrase *equine protozoal myeloencephalitis* appears for the first time in any US patent in patent 5,747,746. Beginning with this example, it goes on to appear in several other patents in animal health. In contrast to *pyrimethamine*, the concept of (treating) *equine protozoal myeloencephalitis* is one that was born in agriculture, reflecting the primarily agricultural research base upon which it is based.

1.3 Main Results

We here present five different measures of knowledge spill-ins to agriculture. We begin with results that use patent citations and then present results that rely on citations to nonpatent literature and results that use shared text concepts.

1.3.1 Patent Citations

In figure 1.3, we show the share of citations made by each agricultural subsector that originate in their own subsector (i.e., animal health patents citing animal health patents) and other subsectors (i.e., animal health patents citing research tool patents).

It is apparent that for the first four agricultural subsectors, more than half of citations accrue to patents not classified as agricultural patents. This indicates a substantial role for knowledge spill-ins from outside agriculture. In these four sectors, the second most cited subsector is the own subsector. There is very little knowledge flow between different agricultural subsectors.

In contrast, the majority of citations in the plants and research tools sub-

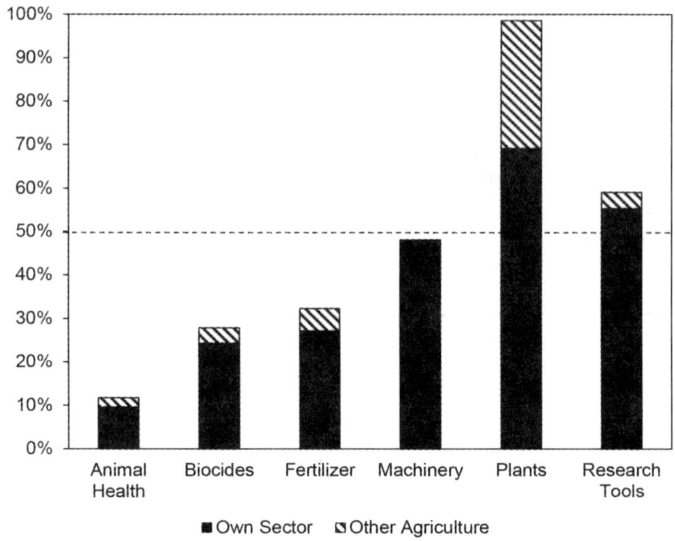

Fig. 1.3 Share of patent citations to agricultural subsectors

Note: The citing sector is on the horizontal axis. Only cited patents granted between 1976 and 2016 are included. Each citation is counted once, even if multiple citations point to the same patent. Own sector gives the share of these citations to patents in the same subsector. Other agriculture gives the share of these citations to any other agricultural subsector. The remaining share of citations accrues to patents not contained in any of our agricultural subsectors.

sectors accrue to patents that belong to these subsectors. While the research tools subsector still cites a substantial number of patents outside of agriculture (40.9 percent), in the plants subsector, citations to other plant patents and to research tool patents account for almost 100 percent of all citations made.

Table 1.6 breaks down the share of citations from each subsector to the type of assignee/inventor associated with the cited patent. As noted in section 1.2.3, we divide nonindividual assignees into four categories: assignees (mostly firms) specializing in agricultural R&D, assignees (mostly firms) that conduct some agricultural R&D but for whom such activities are the minority, assignees (mostly firms) conducting no agricultural R&D, and the public sector (mostly government, universities, and not-for-profit organizations). We omit the patents of unclassified assignees, which never receive more than 1.5 percent of citations.

Only in the plants subsector do the majority of cited patents belong to assignees that specialize in agriculture. A plurality of patent citations in the machinery subsector also originates with assignees that specialize in agriculture. For animal health, biocides, fertilizer, and research tools, either a plurality or a majority of patent citations originate in agricultural minority firms. In no sector do more than 21 percent of patent citations originate with

Table 1.6 **Share of patent citations to assignee types**

	Ag specialized (%)	Ag minority (%)	Non-ag (%)	Public sector (%)	Individuals (%)
Animal health	1.8	69.1	18.4	4.1	6.2
Biocides	8.6	65.1	13.2	4.6	7.8
Fertilizer	17.4	33.7	20.7	4.5	23.5
Machinery	33.5	29.1	8.8	1.1	27.5
Plants	80.6	5.4	0.3	12.8	0.6
Research tools	28.1	38.2	12.8	13.6	5.8

Note: The rows indicate the citing agricultural subsector and columns the assignee and inventor type to which the cited patents belong. Specialized ag assignees have more than 50 percent of their patents belonging to an agricultural subsector in the last five years. Minority ag assignees have more than zero patents belonging to an agricultural subsector in the last five years but less than 50 percent. Non-ag assignees have no patents belonging to agricultural subsectors. Public sector assignees correspond to government and nonprofit organizations. "Individuals" refers to patents owned by individual inventors. Rows do not add up to 100 percent—the remainder of patent citations are made to unclassified assignees (see section 1.2.3.1).

assignees that do not conduct any agricultural research (even though these assignees account for 55 percent of all patents over this period). Public sector research is disproportionately important for all firms (considering that it accounts for just 2 percent of all patents) and especially important for plant and research tools patents.

Figure 1.4 presents more granular information on the agricultural focus of cited patents. For each point (x,y), share y of all citations made by the subsector accrue to patents belonging to firms with x percent or less agricultural patents over the past five years. Note that this sample is conditional on the citation going to an assignee and not a public sector organization or individual inventor.

The concave-to-convex curves in most of these figures tell us that most citations go to firms that either are very specialized in agriculture (i.e., a very large share of the assignees' patents are classified as agricultural) or have only a tiny agricultural R&D operation (i.e., a very small share of the assignee's patents are agricultural). Only the machine subsector is an outlier, with an approximately linear curve. No curve has a convex-to-concave *S* shape, which would characterize the presence of many cited assignees with an agricultural focus near 50 percent. This suggests our division of assignees into agricultural minority and agricultural specialized is a reasonable one. It also suggests most of the agricultural minority patents have only a tiny footprint in agriculture.

1.3.2 Journal Citations

In figure 1.5, we present the share of matched SCImago journal citations belonging to different journal categories.

Only in the plants subsector do the majority of cited journals belong to

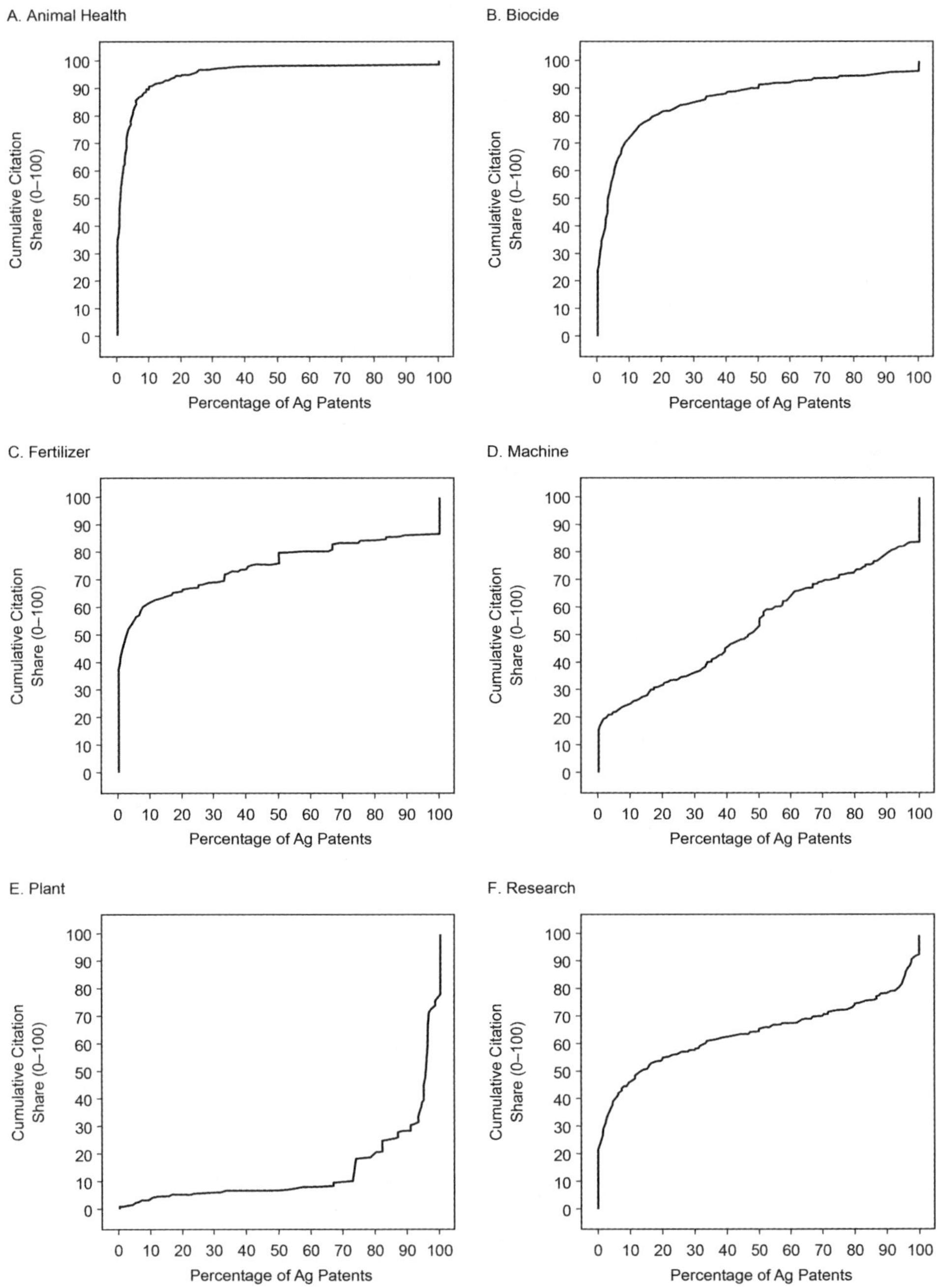

Fig. 1.4 Share of citations to assignees by agriculture specialization

Note: Cumulative distribution function for citations by agricultural focus of cited assignee. For each point (x,y), share y of all citations made by the subsector accrue to patents belonging to firms with x percent or fewer agricultural patents over the past five years. Note that this sample is conditional on the citation going to an assignee and not a public sector organization or individual inventor.

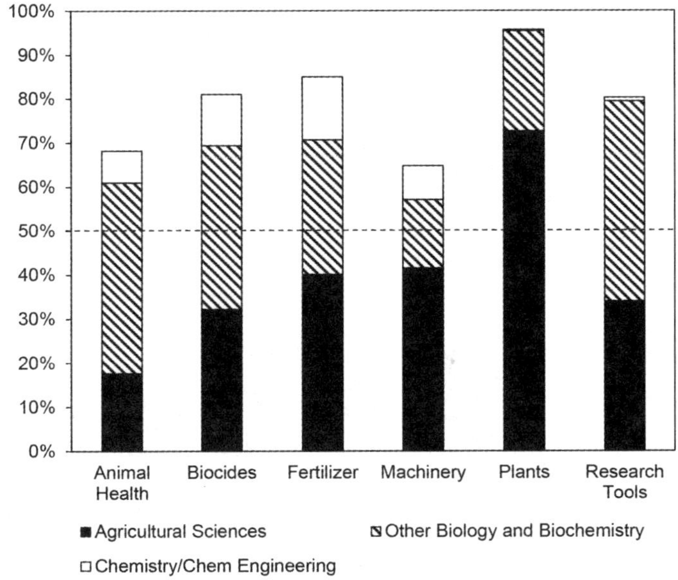

Fig. 1.5 Share of journal citations to journal categories

Note: Citing agricultural subsector is listed on the horizontal axis. Shares are given conditional on matching journal title to the SCImago database. The remaining share of citations to journals accrues to other journals in SCImago that we do not classify as one of the above categories.

the agricultural sciences category. In the fertilizer and machinery subsectors, a plurality of cited journals belong to the agricultural sciences sector. With the exception of machinery, the other biology and biochemistry category is either the most or next-most important category of cited journals. In the machinery subsector, other journals are the second-most important source.

1.3.3 Shared Text Concepts

Our shared text concept results are designed to detect the sources of important new (or at least recently reawakened) concepts in agriculture. An important difference compared to the foregoing analysis is that whereas citations track knowledge flows "one step removed," our text approach can accurately track the "deep roots" of knowledge spill-ins. For example, an idea originating in a distant technology sector may pass through a long sequence of citations before finally being cited by an agricultural patent. To generate the following results, we perform the following calculation for each text-novel concept (see section 1.2.2) in each subsector. First, we identify the earliest subsector patent that mentions the concept. We use the application date of this patent as the date this text-novel concept is first applied in that subsector.

Next, we look for any mention of the concept in patents granted prior to

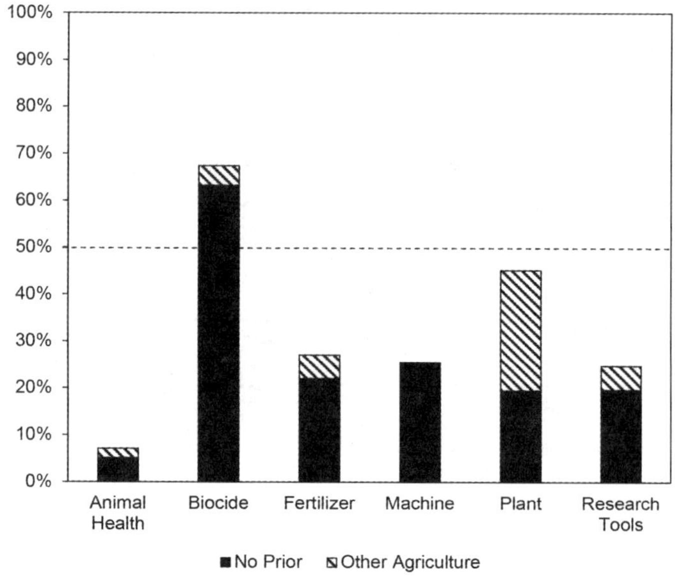

Fig. 1.6 Probability of antecedent text-novel concept mentions across agricultural subsectors

Note: Each bar gives the probability of a randomly selected patent mentioning a randomly mentioned text-novel concept that originates in a given sector. Antecedent mentions are all those made by patents applied for prior to the first patent in the given subsector that mentions the concept. The remaining share of antecedent mentions accrues to patents not classified as agriculture.

this date. By construction, none of these patents will be in the "own subsector" prior to this date, but they may have been used in other agricultural subsectors or outside of agriculture. If there are any antecedent patents mentioning the concept, we compute the share of these that belong to each originating knowledge domain. Denote the share of concept *c*'s prior mentions originating in knowledge domain *i* by $s_i(c)$. If no prior patents mention the concept, we say the concept has no prior mentions ($s_i(c) = 1$, with *i* denoting "no prior mentions"). We then take the average share across all text-novel concepts:

$$(1) \qquad p_i = \frac{1}{n}\sum_{c=1}^{n} s_i(c).$$

Intuitively, the interpretation of p_i is the probability that a randomly selected knowledge flow from a randomly selected text-novel concept *c* originates in sector *i*. Figure 1.6 depicts the probability that a random knowledge flow from a concept originates in agriculture.

In the biocides sector, fully 63 percent of top text-novel concepts appear

Table 1.7 **Share of antecedent text-novel concept mentions across assignee type**

	Ag specialized (%)	Ag minority (%)	Non-ag (%)	Public (%)	Individuals (%)	No prior mention (%)
Animal health	1.2	44.1	31.2	7.0	9.4	5.1
Biocide	3.5	26.3	4.8	0.9	0.3	63.3
Fertilizer	2.5	29.8	29.0	4.3	11.2	22.1
Machine	2.8	16.1	42.3	1.0	11.8	25.5
Plant	10.8	28.7	23.3	10.4	5.9	19.5
Research tools	2.1	25.4	30.3	13.3	7.2	19.7

Note: An entry gives the probability that a randomly selected patent mentioning a randomly selected text-novel concept originates with a given assignee type. Antecedent mentions are all those made by patents applied for prior to the first patent in the given subsector that mentions the concept. Specialized ag assignees have more than 50 percent of their patents belonging to an agricultural subsector in the last five years. Minority ag assignees have more than zero patents belonging to an agricultural subsector in the last five years but less than 50 percent. Non-ag assignees have no patents belonging to agricultural subsectors. Public sector assignees correspond to government and nonprofit organizations. "Individuals" refers to patents owned by individual inventors. "No prior mention" indicates the concept has no prior mentions. Rows do not add up to 100 percent—the remainder of patent mentions (0.1–1.5 percent) are made to unclassified assignees (see section 1.2.3.1).

for the first time in the patent corpus as part of the title, abstract, or claims of a biocide patent. This turns out to be an exception. Other than the biocides sector, the majority of text-novel concepts in each subsector are mentioned in earlier patents. The majority of these are mentioned by patents outside of agriculture. Again, there is little transfer of knowledge from within agriculture, with the exception of the plant subsector, where 20 percent of prior mentions come from the research tools subsector and 5 percent from the biocides subsector.

Table 1.7 performs the same exercise for the type of assignee/inventor. Most text-novel concepts are mentioned before their use in agriculture by patents that do not specialize in agricultural R&D. This is consistent with figure 1.6, which establishes that most text-novel concepts are not mentioned in other agricultural sectors prior to their appearance in a given subsector. A large share of these concepts are mentioned, however, in firms with some agricultural research. The plurality of mentions occurs in minority agricultural assignees in four of the six sectors, whereas the plurality occurs in nonagricultural assignees in the other two (machinery and research tools).

Figure 1.7 again presents more granular information on the agricultural focus of patents mentioning text-novel concepts. Any point (x,y) in figure 1.7 gives the cumulative probability x that a randomly selected knowledge flow containing a randomly selected concept belongs to a patent with agricultural focus y or less. Note that this sample is even more restricted than figure 1.6,

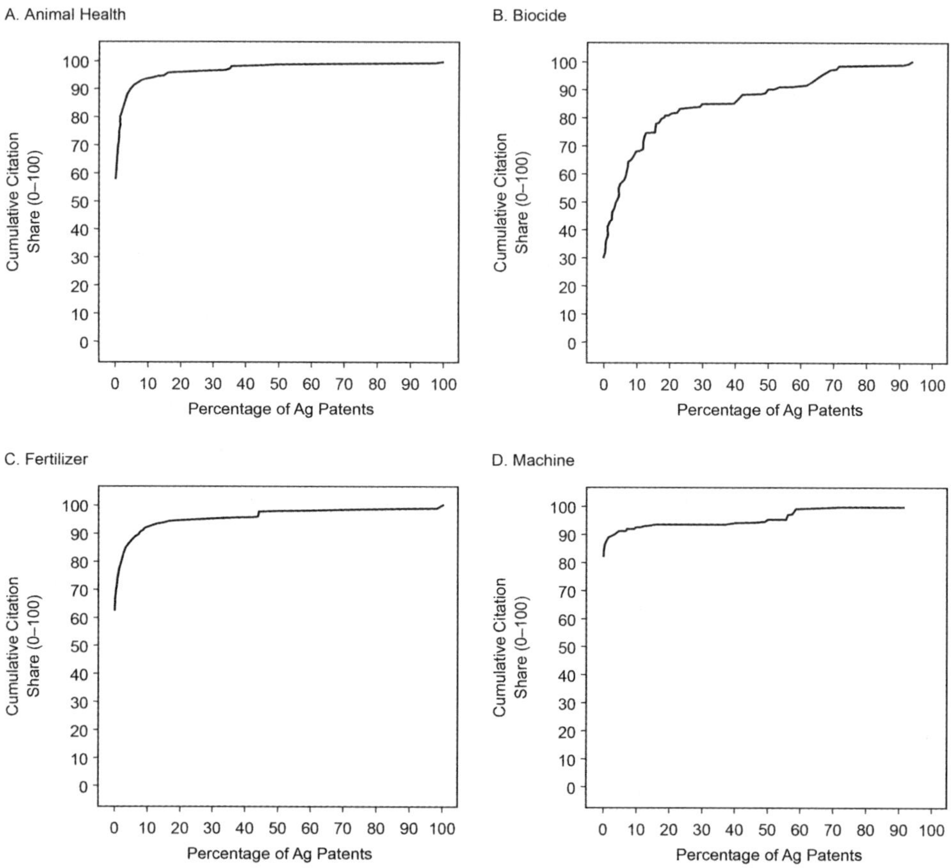

Fig. 1.7 Cumulative probability of antecedent text-novel concept mentions by assignee specialization in agriculture

Note: Cumulative distribution function for prior mentions of text concepts by agricultural focus of cited assignee. Any point (x,y) in figure 1.7 gives the cumulative probability x a randomly selected knowledge flow containing a randomly selected concept belonging to a patent with agricultural focus y or less. Note that this sample is even more restricted than figure 1.6, since it excludes the patents of public sector firms and individuals as well as text concepts that have *no* prior mentions.

since it excludes the patents of public sector firms and individuals as well as text-novel concepts that have *no* prior mentions.

Unlike figure 1.6, these shapes are mostly just concave rather than concave-to-convex (the plant subsector being the only one showing a significant concave ending). This suggests that prior mentions by minority agricultural assignees are mostly assignees with only a small agricultural focus—much less than 50 percent. For important text-novel concepts that are not born in agriculture, they tend to come from firms with either no history in agriculture or a very minor one.

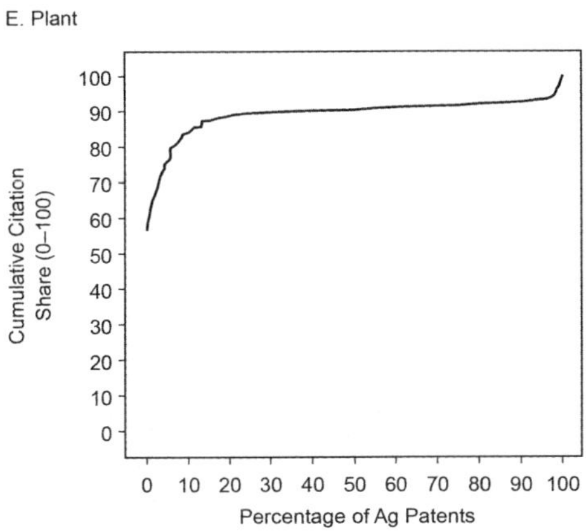

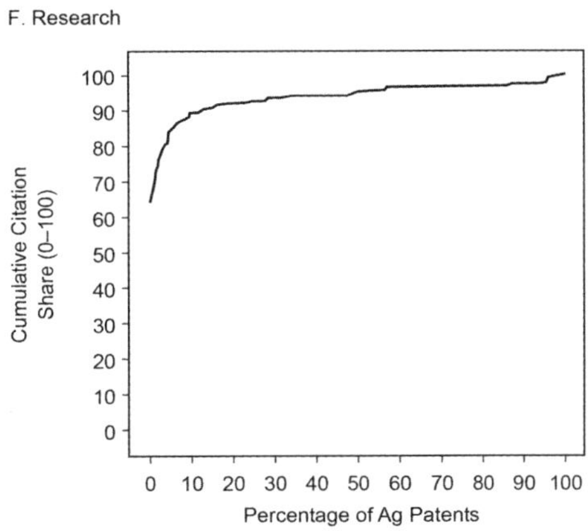

Fig. 1.7 (cont.)

1.4 Discussion

Section 1.3 describes five different measures of the extent of knowledge spill-ins to agriculture. Each measure emphasizes a different potential aspect of spill-ins. Section 1.3.1 emphasizes the flow of knowledge in the space of patented technologies across our entire time period. Section 1.3.3 also focuses on the space of patented technologies but looks specifically at a subset of "concepts" that arose to prominence in agriculture during the second half of our observation period. It measures the extent of prior R&D (potentially many citations removed) related to these concepts outside of the particular agricultural subsector. Section 1.3.2, in contrast, examines the flow of knowledge from the primarily academic sector to patented technology.

Summarizing this heterogenous set of proxies is challenging, but one of our overarching conclusions is that knowledge spill-ins from outside agriculture are likely as important as knowledge generated within agricultural domains. This conclusion is bolstered by figure 1.8, which indicates the share of knowledge flows that originate in an agricultural knowledge domain, defined below.

In this figure, we pull together proxies for the share of knowledge flows originating in agriculture:

- Patent cites 1: Share of patent citations to agricultural subsectors in figure 1.3.
- Patent cites 2: Share of patent citations to specialized agricultural assignees in table 1.6.
- Journal cites: Share of journal citations to agricultural sciences journals in figure 1.5.
- Text concepts 1: Probability that a text-novel concept has either no prior

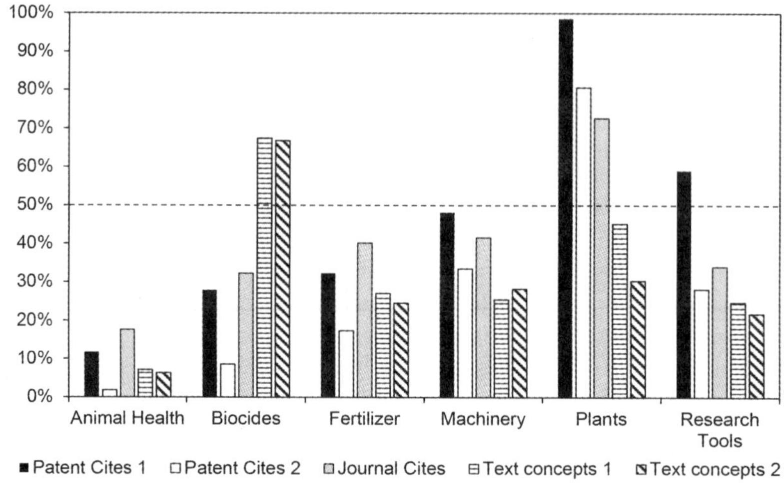

Fig. 1.8 Share of knowledge flows originating within agriculture

Note: Patent cites 1 is the sum of own-sector and other-agriculture bars from figure 1.3. Patent cites 2 is the share of citations going to specialized-ag assignees in table 1.6. Journal cites is the share of journal citations to agricultural science journals from figure 1.5. Text concepts 1 is the sum of no-prior and other-agriculture bars from figure 1.6. Text concepts 2 is the sum of no-prior and ag-specialized categories in table 1.7.

mention or a knowledge flow originating with an agricultural patent in figure 1.6.

- Text concepts 2: Sum of the no prior mention and specialized agricultural columns in table 1.7.

By these definitions, the animal health, fertilizer, and machine subsectors source the majority (more than half) of their ideas from outside agriculture, as measured by any proxy.

The evidence is more mixed for the research tools and biocide subsectors. For research tools, 55 percent of patent citations refer back to other research tools patents, and another 4 percent originate with other agricultural patents. However, most of these patents are assigned to firms that are not specialized in agriculture, and most of the text-novel concepts in research tools patents are mentioned in patents that lie outside agriculture. Moreover, research tools patents cite academic journals at four times the rate of any other sector, but only 34 percent of citations flow to agricultural science journals.

Biocide patent and journal citations primarily flow to nonagricultural firms, patents, and journals. However, the strong majority of text-novel text concepts in biocides have no prior mention and appear for the first time in the patent corpus in a biocide patent. The majority of these concepts are chemical names, suggesting the subsector develops many chemicals for

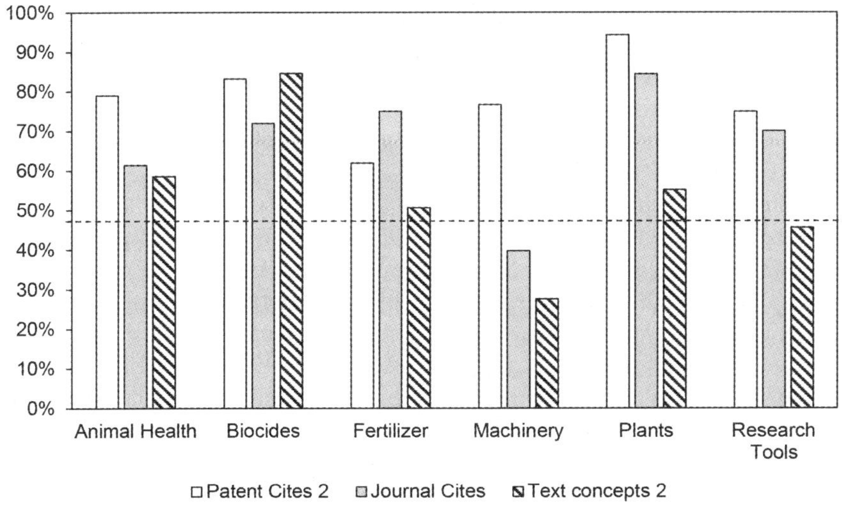

Fig. 1.9 Share of nonagricultural knowledge flows originating "close" to agriculture

Note: Patent cites 2 is the share of citations to nonspecialized ag assignees that are classified as minority ag. Journal cites 2 is the share of nonagricultural journal citations classified as biology/biochemistry or chemical / chemical engineering. Text concepts 2 is the share of prior text mentions by nonspecialized ag assignees that are classified as minority ag.

application in agriculture that appear nowhere else in the patent corpus. This is an observation that would be missed if we relied solely on citations.

Finally, plants seem to be different. The majority of citations flow to specialized agricultural firms, agricultural patents, and agricultural science journals. For text concepts, the majority are mentioned in nonagricultural patents before their appearance in patents for plant varieties, but not by an overwhelming number (55 percent). It is important to note that utility patents for plants differ from other utility patents in more than just their subject matter. This field is dominated to an unusual extent by a small number of firms, with some evidence that they use a standardized template for new patents (Moser, Ohmstedt, and Rhode 2018).

Taken together, in no field do all our knowledge-flow proxies agree that agriculture is the main source of inputs. Rather, spill-ins from outside agriculture appear to matter—and to matter a great deal in most subsectors. We now turn to the nature of these nonagricultural spill-ins.

Whereas our chapter does not try to rigorously define the "distance" between different knowledge domains, our results do provide some evidence that knowledge flows from outside of agriculture do not originate "too far" from agriculture. In figure 1.9, we present an attempt to measure whether knowledge flows originate "far" from agriculture by resorting to some reasonable but perhaps ad hoc assumptions. We assume research originating in "nonagricultural" assignees (tables 1.6 and 1.9) is further from agriculture than research originating in "minority agricultural" firms. This would be the case, for example, if an assignee's knowledge capital has some agricultural applications. In this case, the fact that the assignee also patents in agriculture is a signal that it has recognized the agricultural application of its knowledge capital. The animal health sector would seem to be a good example of this kind of dynamic. Much of the basic research on health for humans or animals is similar at the cellular level, even though the human health market is vastly larger than the veterinary health market (Sneeringer and Clancy 2020). That said, caution is warranted, because an assignee may also be a conglomerate with many parallel research operations that effectively embody separate knowledge capital stocks.

We feel it is also reasonable to assume biology and chemistry are scientific disciplines that are among the closest to agriculture, and so citations to biological and chemistry journals are indicators that fields "close" to agriculture matter. Agriculture is typically classified as one of the life sciences (e.g., by the National Science Foundation [NSF]), and agricultural science has deep roots in chemistry (Huffman and Evenson 2006). Figure 1.9 uses these notions to provide some evidence that knowledge from outside agriculture is not "too far" away.

In this figure, we pull together very rough proxies for the distance from agriculture of nonagricultural knowledge flows.

- Patent cites 2: Share of citations to assignees, but not specialized agricultural assignees, that are classified as minority agricultural (as opposed to nonagricultural).
- Journal cites: Share of nonagricultural science journal citations to journals classified as biology/biochemistry or chemical/chemical engineering (as opposed to "other").
- Text concepts 2: Share of prior text mentions by assignees, but not specialized agricultural assignees, that are classified as minority agricultural (as opposed to nonagricultural).

In contrast to figure 1.9, most proxies now clear the 50 percent line. Where we can reasonably rank knowledge domains as being closer or further from agriculture, nonagricultural knowledge flows in animal health, biocides, fertilizer, and plants are more likely to come from knowledge domains close to agriculture than from afar. For machinery and research tools, text concepts tend to be mentioned more often in nonagricultural assignees than in minority agricultural ones. Machinery is also more likely to cite other journals than biology or chemistry ones, which is not surprising. Note, however, that the machinery sector cites by far the fewest journal publications.

Together, figures 1.8 and 1.9 suggest that while nonagricultural knowledge sources are very important, some nonagricultural knowledge domains are clearly more relevant than others. Whereas we view this conclusion as more tentative than our first one, it has relevance for science policy in agriculture.

1.5 Robustness Checks

In this section, we conduct a wide array of robustness checks. To prevent the main chapter from becoming too long, we report tables in the appendix and merely summarize important details in the text.

1.5.1 Patent Citations

We investigate three potential sources of bias in our patent citation figures: first, that our results are driven by assignee's self-citation of their own patents; second, that our results are robust to the exclusion of examiner-added citations; and third, that our results are robust when we restrict attention only to the most valuable patents (those receiving a high number of citations themselves).

There is a debate about the extent to which patent citations may be biased by a tendency for firms to cite their own work or by the additional citations added by patent examiners (Lampe 2012; Moser, Ohmstedt, and Rhode 2018). To assess whether our results are driven by self-citation, we first remove all citations from assignees to their own patents. Because so many individual inventors have a single patent, and because it is harder to accurately disambiguate inventor names, we restrict attention to assignee self-citation. The results are presented in tables 1.A2 and 1.A3.

Excluding self-citations does not materially change the distribution of patent citations across different agricultural sectors, with one exception. In figure 1.3, the share of citations from plant patents to plant patents is 69 percent, but when we exclude self-citations, this falls to 56 percent. Similarly, in table 1.6, the share of citations to specialized agricultural firms is 81 percent, but when we exclude self-citations, this falls to 69 percent. Moser, Ohmstedt, and Rhode (2018), studying a sample of hybrid corn patents granted between 1985 and 2002, find that self-citations frequently reflect genuine cumulative innovation, as firms build on the prior genetic stock of their earlier patented plant cultivars. Therefore, it is not at all clear that the smaller share of 56 percent should be preferred to our baseline estimate of 69 percent.

Next, we remove all examiner-added citations. This is only possible for the period 2002 onward, when patents begin to identify who added a citation. There is some debate about whether examiner-added citations are good proxies for knowledge flows. If applicants seek to avoid citing relevant prior art for strategic reasons, examiner-added citations can correct this bias (Lampe 2012). Moreover, Chen (2017) finds examiner-added citations are more textually similar to the patent than other patents. That said, there is a large literature that highlights potential issues with examiner-added citations. For example, Moser, Ohmstedt, and Rhode (2018) find that examiners of hybrid corn patents are biased toward adding from their set of preferred patents and that patents will tend to be added more for physical similarity of plants rather than genetic heritage. Jaffe and de Rassenfosse (2017) summarize a number of other studies that describe potential distortions examiner-added citations may introduce. Tables 1.A4 and 1.A5 present the distribution of patent citations for patents granted after 2002, excluding examiner-added citations.

Removing examiner-added citations leaves our results largely unchanged, with one exception. In the machinery subsector, in figure 1.3 we found that 48 percent of patents citations originated in the machinery subsector and 52 percent originated outside of agriculture. In table 1.A4, we instead find that 56 percent of citations originate in the machinery subsector and 44 percent originate outside of agriculture. It turns out, however, that this has little to do with examiners and is instead driven by restricting patents to those granted after 2002. If we restrict attention to patents granted after 2002 (table 1.A6), 56 percent of patent citations in the machinery subsector originate in the same sector. Indeed, across all subsectors, there is a slight increase in patents originating from within the same subsector when we restrict attention to more recent patents.

Our final robustness check relates to the heterogenous value of patents. Many studies (see Nagaoka, Motohashi, and Goto 2010 for an overview) have shown that the value of patents is highly skewed. A small number of patents account for a disproportionately large share of value. Our results may be misleading if the minority of valuable patents differ in the sources of their knowledge compared to patents as a whole. To check this, we identify the set of most valuable patents in agriculture, defined as those receiving eight or more citations in the five years following the date they are granted (this necessarily means we do not include patents from the last five years of our sample).[9] Patents receiving eight or more citations are in the top 5 percent for all agricultural patents. Tables 1.A7 and 1.A8 repeat our patent citation analysis for this subset of elite patents.

Restricting our attention to only the citations made by "elite" patents, we find that a significantly higher share of citations originate from within the same subsector for the fertilizer, machinery, and research tools subsectors. Indeed, for machinery, the effect is large enough to tip the share of citations originating in the machinery subsector above 50 percent, from 48 percent in figure 1.3 to 64 percent, in table 1.A7. In no other sector, however, does the share of citations from a given sector cross the 50 percent threshold, and so the conclusions drawn from our figures 1.8 and 1.9 remain valid. Turning to the share of citations received by different assignee types, restricting attention to only the most highly cited patents has the largest impact for the plant subsector, where the share of citations to specialized agricultural firms drops from 81 percent to 67 percent, and the share of citations to public sector patents rises from 13 percent to 25 percent.

9. Citations received is a common proxy for the value of patents. See Nagaoka, Motohashi, and Goto (2010).

1.5.2 Journal Citations

Flagship multidisciplinary journals such as *Science*, *Nature*, and *PNAS* present a challenge to our journal citation analysis. We classify these journals as "other," but citations to these journals could conceivably be to top articles in agricultural science, biology, or chemistry. In table 1.A9, we break out citations to these three journals as a separate category. In the research tools subsector, these three journals account for 14.2 percent of all journal citations. However, even if the cited articles are all agricultural science articles, we still find about 50 percent of all journal citations would be to agricultural science. In the other subsectors, these three journals account for 1.3–3.9 percent of citations to academic work, suggesting that the main conclusions presented in section 1.4 are robust.

1.5.3 Text Concepts

We check the robustness of our text concept analysis to three alternative specifications. First, we impose stricter criteria to our manual cleaning of concepts in agriculture. Second, we use an alternative weighting scheme that controls for the possibility that some of our concepts are duplicates that refer to the same underlying idea. Third, we use an alternative weighting scheme that puts more weight on clusters of concepts that are used in more future patents.

Tables 1.A10 and 1.A11 impose stricter criteria to our manual cleaning of text-novel concepts in agriculture. To manually clean concepts, three coauthors independently apply four exclusion rules (see section 1.2.2) to all concepts in our data. There is some subjectivity in these rules—for example, in judging what is an "uninformative" word and what "connective phrases" are. In the main specification, we retain a concept when at least two of the three judges retain it. In our robustness check, we require all three inspectors to agree for a concept to be retained. Depending on the subsector, this leads to us excluding an additional 10–20 percent of the original 200 concepts (the set of included and excluded concepts is available in tables A.16–A.21 of the appendix). Our core results, however, are not substantively changed by this stricter exclusion policy. In figures 1.8 and 1.9, none of the bars flip from being above 50 percent to below it, or vice versa.

Tables 1.A12 and 1.A13 summarize our text data in a different way. One possible concern with our text-analysis approach is that we may be "double counting" some concepts. This could occur, for example, if two concepts refer to the same underlying idea. For example, suppose pyrimethamine is exclusively used to treat variants of the disease myeloencephalitis. Whenever the concept pyrimethamine appears in a patent, so too does the phrase *myeloencephalitis*, and vice versa, although perhaps not in the same sentence (or paragraph). Section 1.3.3 treats these two phrases as distinct concepts. There, we compute the share of prior mentions for each of these concepts and then average over all these shares. But it could be argued the two concepts pyrimethamine and myeloencephalitis only really refer to one underlying idea (treating the disease with the antibiotic), since they are always and everywhere used together. If this is correct, then we are giving too much weight to the shares of prior patents mentioning these concepts by counting each concept separately.

Here, we consider an alternative approach that creates "families" of related concepts. For each concept, we look for its first appearance in a given agricultural subsector, which we call an originating patent. All concepts in the same originating patent constitute a family of related concepts.

For example, if *pyrimethamine* and *myeloencephalitis* are always used together, then they will both appear for the first time in animal health in the same patent and therefore will belong to the same family. For each of these families, we find the set of unique patents applied for before the originating patent with *any* concepts in the family. We compute the share of these patents originating in different knowledge domains. Denote the share of patents with concepts from family f that originate in knowledge domain i by $s_i(f)$.

We then average these shares over all families:

$$p_i = \frac{1}{n}\sum_{f=1}^{n} s_i(f). \tag{2}$$

This methodology uses originating patents to define families of related concepts and give each family the same weight, ensuring we do not double count concepts referring to the same concept. The trade-off with this approach is that a concept with no prior mentions may belong to a family of concepts that do have prior mentions. This methodology obscures the fact because it treats families of concepts as units of observation.

This alternative methodology does have some significant impacts on our results, but none large enough to alter the conclusions in figures 1.8 and 1.9. Indeed, our major conclusion that ideas from outside of agriculture are important is actually strengthened. Under this alternative weighting scheme, the share of concepts originating in patents outside agriculture rises in every subsector, as does the share of concepts originating in the patents of nonagricultural assignees.

Lastly, we weight families of concepts by the number of agricultural patents that end up using any concepts in the family. Let $w(f)$ denote the number of patents in a subsector that use any concept in family f. Our final weighting scheme is as follows:

$$p_i = \frac{\sum_{f=1}^{n} w(f)s_i(f)}{\sum_{f=1}^{n} w(f)}. \tag{3}$$

Intuitively, this puts more weight on families of concepts that subsequently end up being used more heavily in the agricultural subsector. The results, presented in tables 1.A14 and 1.A15, do not differ materially from tables 1.A12 and 1.A13, although they again tend to increase the weight put on families of concepts originating outside of agriculture.

1.6 Conclusions

Agricultural total factor productivity grew enormously over the past century. In the years to come, continued increases in agricultural productivity will be essential for meeting the challenge of feeding a rising world population amid the trials of climate change. There is widespread recognition that past R&D investments were crucial to developing the new and improved agricultural technologies that have mediated these celebrated productivity gains. This chapter presents new evidence on the structure of knowledge underpinning agricultural R&D, with an emphasis on the role of knowledge spillovers across scientific and technological domains.

Using agricultural patents in animal health, biocides, fertilizer, machinery, plants, and research tools as measures of agricultural research outputs, we track knowledge flows into agriculture in five different ways. We start with citations to patents in agricultural subsectors and across different types of inventive organizations and individuals. To capture knowledge flows from academia, we also track citations to journal articles across different journal categories. Finally, we complement these citation-based approaches with text analysis, where we identify text concepts that are new (in text) and important in agriculture in the second half of our observation period. We then track the appearance of these text concepts in earlier patents.

Our results indicate a major role for ideas that originate outside of agriculture, perhaps a role as important as R&D conducted within agriculture. In the animal health, fertilizer, and machinery subsectors, across every measure we find that the majority of knowledge flows originate in nonagricultural knowledge domains. In the remaining three subsectors, we find mixed evidence: some of our indicators suggest that the majority of knowledge originates outside agriculture, while some originates from within. Amid these sets, the strongest case for knowledge originating primarily from within agriculture is the plant subsector, which mainly cites other agricultural patents and agricultural science journals. But even this subsector has the majority of its text concepts appearing outside of agriculture prior to their appearance in plant patents.

We also present some evidence that these "outside agriculture" knowledge domains remain predictably close to agriculture. Whereas agricultural science journals do not account for the majority of journal citations in most subsectors, together with biology and chemistry journals, they do. Moreover, our other measures of knowledge flows indicate that organizations with at least some agricultural patents do R&D more relevant to agriculture than organizations with no agricultural patents.

The novelty of this chapter is to use information contained in patents, through patent citations and text analysis, to study agricultural knowledge flows, and this work suggests a number of possible avenues for future research. First, our text concept approach can be easily extended to the corpus outside of patents. In particular, academic journals are a promising avenue to explore. For example, we find the biocide sector originates the majority of its text concepts and that these concepts tend to be chemical names. At the same time, the sector heavily cites chemistry journals, and it would be interesting to see if these chemical names appear first in chemistry journals. Outside of agriculture, Li, Azoulay, and Sampat (2017) track knowledge flows from life science patents to basic research and find that 30 percent of National Institutes of Health (NIH) grants result in publications that are subsequently cited by life science patents. Our text-analysis approach could help identify cases where NIH funding results in ideas that are used in life science patents without a direct citation. More generally, this approach can be extended to books, company filings, and so on.

Second, the combination of text-novel concepts and citations represents a clear opportunity to track the diffusion of specific ideas through technology space. Are citations a channel through which text concepts flow, and if so, can we track the movement of an idea originating in one technology field through a chain of linked citations to an eventual application in a distant technology field? This would allow one to examine the factors that most facilitate the transfer of ideas. Lastly, the analysis we have presented can be brought to bear on work linking agricultural R&D to agricultural productivity measures. Patents may serve as new proxies for knowledge capital—proxies with more detailed information about the relevant R&D spending both in agriculture and beyond.

Albeit preliminary, we may attempt to draw some normative implications of the results presented in this chapter. The early work of Schultz (1956) and Griliches (1958) underscored agriculture's leading position in identifying the role of technical progress on productivity. A large and varied literature has since established the fundamental role that investments in science and technological R&D have on innovation and economic growth. The many market failures that beset the innovation process suggest a critical role for public policies to fund and support the R&D enterprise. Evidence of past remarkable successes has fostered the belief that scientific research is underfunded and that a renewed investment impetus would raise economic growth. The insight is particularly relevant for US agriculture, where public R&D investments have substantially declined, in real terms, over the last decade (USDA ERS 2019). Meritorious calls for increased public agricultural R&D inevitably meet the reality of the declining availability of public funds. In this age of scarcity, science policy needs to be mindful of the complexity and connectedness of the research enterprise. As highlighted in the model of Akcigit, Hanley, and Serrano-Velarde (2021), the spillover effects from basic research are critical. In our context, the knowledge spillovers we have identified suggest that agricultural science policy might best support agricultural productivity growth if it retains a holistic perspective. Attention to the broader research agenda—and in particular, to areas that while not being strictly agriculture oriented have traditionally been connected with agricultural innovation—is of paramount importance. Priorities that rely on narrowly defined measures of past returns to R&D may not provide the most productive use of scarce public R&D funds.

Appendix

Table 1.A1 **Share of patents derived from domestic and foreign research**

	Derived from	
	Domestic research (%)	Foreign research (%)
Animal health	65.4	34.6
Biocides	51.2	48.8
Fertilizer	59.8	40.2
Machinery	64.5	35.5
Plants	85.9	14.1
Research tools	59.7	40.3

Note: Patents are fractionally classified as derived from domestic or foreign research based on the share of inventors listing a US (domestic) or non-US (foreign) address.

Table 1.A2 **Share of patent citations to agricultural subsectors, excluding assignee self-citations**

	Own sector (%)	Other agriculture (%)	Not agriculture (%)
Animal health	5.8	2.4	91.8
Biocides	23.1	3.6	73.3
Fertilizer	26.3	5.0	68.7
Machinery	46.3	0.1	53.6
Plants	56.1	41.8	2.1
Research tools	53.3	3.5	43.2

Note: The rows indicate the citing agricultural subsector and columns the subsector to which cited patents belong. Only cited patents granted between 1976 and 2016 are included. Each citation is counted once, even if multiple citations point to the same patent. "Own sector" gives the share of these citations to patents in the same subsector. "Other agriculture" gives the share of these citations to any other agricultural subsector. "Not agriculture" gives the share of citations to patents not contained in any of our agricultural subsectors. We exclude citations made by assignees to their own patents.

Table 1.A3 **Share of patent citations to assignee types, excluding self-citations**

	Ag specialized (%)	Ag minority (%)	Non-ag (%)	Public sector (%)	Individuals (%)
Animal health	1.8	63.1	22.2	4.8	7.5
Biocides	8.5	62.5	14.8	4.6	8.8
Fertilizer	17.0	32.0	21.7	4.4	24.6
Machinery	32.0	27.7	9.5	1.1	29.6
Plants	69.2	8.5	0.5	20.4	1.0
Research tools	26.1	37.9	14.1	14.0	6.4

Note: The rows indicate the citing agricultural subsector and columns the assignee and inventor type to which the cited patents belong. Specialized ag assignees have more than 50 percent of their patents belonging to an agricultural subsector in the last five years. Minority ag assignees have more than zero patents belonging to an agricultural subsector in the last five years but less than 50 percent. Non-ag assignees have no patents belonging to agricultural subsectors. Public sector assignees correspond to government and nonprofit organizations. Individuals refers to patents owned by individual inventors. Rows do not add up to 100 percent—the remainder of patent citations (0.1–1.4 percent) are made to unclassified assignees (see section 1.2.3.1). We exclude citations made by assignees to their own patents.

Table 1.A4 **Share of patent citations to agricultural subsectors (2002 and later), excluding examiner-added citations**

	Own sector (%)	Other agriculture (%)	Not agriculture (%)
Animal health	6.9	2.4	90.7
Biocides	24.4	4.7	70.8
Fertilizer	29.3	6.6	64.1
Machinery	56.4	0.2	43.5
Plants	67.0	31.8	1.2
Research tools	55.9	3.3	40.9

Note: The rows indicate the citing agricultural subsector and columns the subsector to which cited patents belong. Only cited patents granted between 1976 and 2016 are included, and only citing patents granted after 2002 are presented. Each citation is counted once, even if multiple citations point to the same patent. "Own sector" gives the share of these citations to patents in the same subsector. "Other agriculture" gives the share of these citations to any other agricultural subsector. "Not agriculture" gives the share of citations to patents not contained in any of our agricultural subsectors. We exclude citations made by patent examiners.

Table 1.A5	Share of patent citations to assignee types (2002 and later), excluding examiner-added citations				
	Ag specialized (%)	Ag minority (%)	Non-ag (%)	Public sector (%)	Individuals (%)
Animal health	1.2	64.0	25.0	4.7	4.9
Biocides	9.3	62.6	14.9	4.4	7.7
Fertilizer	18.0	31.3	23.1	5.4	22.0
Machinery	35.7	28.5	9.5	1.3	25.0
Plants	79.4	5.5	0.3	14.0	0.6
Research tools	28.0	38.7	13.5	13.2	5.2

Note: The rows indicate the citing agricultural subsector and columns the assignee and inventor type to which the cited patents belong. Only cited patents granted between 1976 and 2016 are included, and only citing patents granted after 2002 are presented. Specialized ag assignees have more than 50 percent of their patents belonging to an agricultural subsector in the last five years. Minority ag assignees have more than zero patents belonging to an agricultural subsector in the last five years but less than 50 percent. Non-ag assignees have no patents belonging to agricultural subsectors. Public sector assignees correspond to government and nonprofit organizations. Individuals refers to patents owned by individual inventors. Rows do not add up to 100 percent—the remainder of patent citations (0.1–1.4 percent) are made to unclassified assignees (see section 1.2.3.1). We exclude citations made by patent examiners.

Table 1.A6	Share of patent citations to agricultural subsectors (2002 and later)		
	Own sector (%)	Other agriculture (%)	Not agriculture (%)
Animal health	8.7	2.5	88.8
Biocides	26.0	4.5	69.5
Fertilizer	31.0	6.2	62.8
Machinery	55.7	0.1	44.1
Plants	69.9	28.8	1.2
Research tools	57.0	3.5	39.5

Note: The rows indicate the citing agricultural subsector and columns the subsector to which cited patents belong. Only cited patents granted between 1976 and 2016 are included, and only citing patents granted after 2002 are presented. Each citation is counted once, even if multiple citations point to the same patent. "Own sector" gives the share of these citations to patents in the same subsector. "Other agriculture" gives the share of these citations to any other agricultural subsector. "Not agriculture" gives the share of citations to patents not contained in any of our agricultural subsectors.

Table 1.A7 **Share of patent citations from highly cited patents to agricultural subsectors**

	Own sector (%)	Other agriculture (%)	Not agriculture (%)
Animal health	0.0	0.0	100.0
Biocides	25.8	7.0	67.2
Fertilizer	41.1	1.9	57.0
Machinery	63.7	0.1	36.2
Plants	61.3	37.2	1.5
Research tools	68.1	2.2	29.7

Note: The rows indicate the citing agricultural subsector and columns the subsector to which cited patents belong. Only cited patents granted between 1976 and 2016 are included and only citing patents that receive eight or more citations in the five years after their grant dates. Each citation is counted once, even if multiple citations point to the same patent. "Own sector" gives the share of these citations to patents in the same subsector. "Other agriculture" gives the share of these citations to any other agricultural subsector. "Not agriculture" gives the share of citations to patents not contained in any of our agricultural subsectors.

Table 1.A8 **Share of patent citations from highly cited patents to assignee types**

	Ag specialized (%)	Ag minority (%)	Non-ag (%)	Public sector (%)	Individuals (%)
Animal health	0.0	88.9	11.1	0.0	0.0
Biocides	10.3	72.7	9.1	1.9	5.1
Fertilizer	23.9	30.2	24.3	2.4	19.1
Machinery	42.1	27.4	5.3	1.1	24.0
Plants	67.5	6.8	0.2	24.9	0.5
Research tools	30.7	47.8	7.8	8.5	4.0

Note: The rows indicate the citing agricultural subsector and columns the assignee and inventor type to which the cited patents belong. Only cited patents granted between 1976 and 2016 are included and only citing patents receiving eight or more citations within the first five years after being granted. Specialized ag assignees have more than 50 percent of their patents belonging to an agricultural subsector in the last five years. Minority ag assignees have more than zero patents belonging to an agricultural subsector in the last five years but less than 50 percent. Non-ag assignees have no patents belonging to agricultural subsectors. Public sector assignees correspond to government and nonprofit organizations. Individuals refers to patents owned by individual inventors. Rows do not add up to 100 percent—the remainder of patent citations (up to 1.1 percent) are made to unclassified assignees (see section 1.2.3.1).

Table 1.A9 **Share of journal citations to journal categories, separating out science, nature, and PNAS**

	Agricultural sciences (%)	Other biology and biochemistry (%)	Chemistry/ chemical engineering (%)	Science, nature, PNAS (%)	Other Scimago (%)
Animal health	17.6	43.4	7.2	3.4	28.3
Biocides	32.3	37.2	11.6	3.9	15.1
Fertilizer	40.1	30.6	14.3	1.3	13.7
Machinery	41.6	15.6	7.7	2.7	32.5
Plants	72.7	22.8	0.3	3.8	0.5
Research tools	34.0	45.4	0.8	14.2	5.6

Note: Each entry is the share of identified journal citations originating in patents in the row subsector that go to journals in the column category.

Table 1.A10 **Share of antecedent text-novel concept mentions to agricultural subsectors, strict inclusion criteria**

	No prior mention (%)	Other agriculture (%)	Not agriculture (%)
Animal health	4.9	2.0	93.1
Biocide	65.7	4.3	30.0
Fertilizer	20.2	4.2	75.6
Machine	32.9	0.0	67.1
Plant	17.0	28.8	54.2
Research tools	23.8	5.4	70.8

Note: An entry gives the probability that a randomly selected patent mentioning a randomly mentioned text-novel concept originates in a given sector. Antecedent mentions are all those made by patents applied for prior to the first patent in the given subsector that mentions the concept. This table includes a concept only if it is included by all three coauthor inspectors.

Table 1.A11 **Share of antecedent text-novel concept mentions to assignee type, strict inclusion criterion**

	Ag specialized (%)	Ag minority (%)	Non-ag (%)	Public (%)	Individuals (%)	No prior mention (%)
Animal health	1.2	46.0	29.9	7.4	9.1	4.9
Biocide	3.7	25.2	3.7	0.6	0.1	65.7
Fertilizer	2.8	30.4	29.4	4.9	11.3	20.2
Machine	1.3	16.8	35.4	0.9	12.0	32.9
Plant	12.9	31.1	21.4	10.7	5.3	17.0
Research tools	1.9	25.9	27.4	12.7	6.8	23.8

Note: An entry gives the probability that a randomly selected patent mentioning a randomly selected text-novel concept originates with a given assignee type. Antecedent mentions are all those made by patents applied for prior to the first patent in the given subsector that mentions the concept. Specialized ag assignees have more than 50 percent of their patents belonging to an agricultural subsector in the last five years. Minority ag assignees have more than zero patents belonging to an agricultural subsector in the last five years but less than 50 percent. Non-ag assignees have no patents belonging to agricultural subsectors. Public sector assignees correspond to government and nonprofit organizations. Individuals refers to patents owned by individual inventors. "No prior mention" indicates the concept has no prior mentions. Rows do not add up to 100 percent—the remainder of patent mentions (up to 0.8 percent) are made to unclassified assignees (see section 1.2.3.1). This table includes a concept only if it is included by all three coauthor inspectors.

Table 1.A12 **Share of antecedent text-novel concept mentions to agricultural subsectors, weighted by concept family**

	No prior mention (%)	Other agriculture (%)	Not agriculture (%)
Animal health	2.5	3.7	93.8
Biocide	56.8	6.5	36.7
Fertilizer	4.2	7.7	88.1
Machine	16.7	0.1	83.3
Plant	9.4	27.4	63.2
Research tools	17.6	4.5	77.9

Note: An entry gives the probability that a randomly selected patent mentioning a text-novel concept from a randomly selected family of concepts originates in a given sector. Antecedent mentions are all those made by patents applied for prior to the first patent in the given subsector that mentions the concept.

Table 1.A13 **Share of antecedent text-novel concept mentions to assignee-type, weighted by concept family**

	Ag specialized (%)	Ag minority (%)	Non-ag (%)	Public (%)	Individuals (%)	No prior mention (%)
Animal health	0.9	44.3	34.2	5.3	10.3	2.5
Biocide	5.3	29.6	6.2	1.1	0.5	56.8
Fertilizer	1.8	38.4	36.3	5.3	12.8	4.2
Machine	3.3	17.6	46.5	1.2	14.2	16.7
Plant	9.6	33.2	28.5	9.4	7.5	9.4
Research tools	3.2	24.0	31.4	13.1	8.4	17.6

Note: An entry gives the probability that a randomly selected patent mentioning a text-novel concept from a randomly selected concept family originates with a given assignee type. Antecedent mentions are all those made by patents applied for prior to the first patent in the given subsector that mentions the concept. Specialized ag assignees have more than 50 percent of their patents belonging to an agricultural subsector in the last five years. Minority ag assignees have more than zero patents belonging to an agricultural subsector in the last five years but less than 50 percent. Non-ag assignees have no patents belonging to agricultural subsectors. Public sector assignees correspond to government and nonprofit organizations. Individuals refers to patents owned by individual inventors. "No prior mention" indicates the concept has no prior mentions. Rows do not add up to 100 percent—the remainder of patent mentions (up to 1.6 percent) are made to unclassified assignees (see section 1.2.3.1).

Table 1.A14 **Share of antecedent text-novel concept mentions to agricultural subsectors, weighted by concept family and subsequent patents**

	No prior mention (%)	Other agriculture (%)	Not agriculture (%)
Animal health	1.6	3.1	95.3
Biocide	52.3	8.1	39.6
Fertilizer	3.8	7.4	88.9
Machine	14.5	0.0	85.5
Plant	4.9	21.4	73.7
Research tools	14.4	4.2	81.4

Note: An entry gives the probability that a randomly selected patent mentioning a text-novel concept from a randomly selected family of concepts originates in a given sector, where the probability of selecting a concept family is weighted by the number of ag subsector patents using concepts belonging to the family. Antecedent mentions are all those made by patents applied for prior to the first patent in the given subsector that mentions the concept.

Table 1.A15 **Share of antecedent text-novel concept mentions to assignee type, weighted by concept family and subsequent patents**

	Ag specialized (%)	Ag minority (%)	Non-ag (%)	Public (%)	Individuals (%)	No prior mention (%)
Animal health	0.8	44.8	34.5	5.1	10.8	1.6
Biocide	5.9	32.4	6.8	1.1	0.6	52.3
Fertilizer	1.8	38.9	36.6	5.0	12.2	3.8
Machine	4.1	17.2	48.2	1.3	14.0	14.5
Plant	7.8	35.4	32.4	9.0	8.6	4.9
Research tools	3.1	22.6	33.0	15.2	8.4	14.4

Note: An entry gives the probability that a randomly selected patent mentioning a text-novel concept from a randomly selected concept family originates with a given assignee type, where the probability of selecting a concept family is weighted by the number of ag subsector patents using concepts belonging to the family. Antecedent mentions are all those made by patents applied for prior to the first patent in the given subsector that mentions the concept. Specialized ag assignees have more than 50 percent of their patents belonging to an agricultural subsector in the last five years. Minority ag assignees have more than zero patents belonging to an agricultural subsector in the last five years but less than 50 percent. Non-ag assignees have no patents belonging to agricultural subsectors. Public sector assignees correspond to government and nonprofit organizations. Individuals refers to patents owned by individual inventors. "No prior mention" indicates the concept has no prior mentions. Rows do not add up to 100 percent—the remainder of patent mentions (up to 1.6 percent) are made to unclassified assignees (see section 1.2.3.1).

Table 1.A16 Top 117 text-novel animal health concepts

	Unanimous (included in robustness check)	
protozoal	sarcocystis	physiologically active
protozoal myeloencephalitis	cyclooxygenase-2	volatile liquid
equine protozoal myeloencephalitis	milbemycin oxime	bovine respiratory
myeloencephalitis	releasing hormone	bovine respiratory disease
equine protozoal	gonadotropin releasing hormone	respiratory disease
trimethoprim	gonadotropin	swine respiratory
microbial	oral mucosa	pharmacologically active compound
microbial infection	equimolar	pharmacologically active
ear	propofol	diabetes
preservative	prodrug	weaning
terbinafine	cyclooxygenase	without heat
penetration enhancer	bacterial protozoa	heat detection
dermal penetration	bacterial protozoa infections	without heat detection
dermal	protozoa infections	sow
dermal penetration enhancer	kinases	c1-c6alkyl
kinase	polymorph	isoxazoline-substituted
janus kinase	transmucosal	insemination
janus	felis	buprenorphine
bird	ctenocephalides felis	spinosad
injection site	ctenocephalides	linoleic
single injection	hydrate	linoleic acid
hydrophilic surfactant	octyl	animal selected
stearoyl groups	octyl salicylate	catalytic
independently stearoyl groups	succinic	alkyl substituted
stearoyl	succinic acid	tobramycin
palmitoyl groups	xanthan gum	hydroxypropylcellulose
palmitoyl	xanthan	folic acid
asthma	equines	folic
fentanyl	furoate	ethanesulfonic
hydroxypropylmethylcellulose	mometasone furoate	ethanesulfonic acid
hydroxypropylmethylcellulose dissolved	mometasone	methanesulfonic acid
pyrimethamine	gnrh	methanesulfonic
epm	buccal	hydroxypropyl cellulose
prophylactic	cox-2	phenol
	Consensus (excluded in robustness check)	
mediated	containing hydroxypropylmethylcellulose	thickener
independently stearoyl	gonadotropin releasing	breeding
pharmaceutically active agent	synchronizing	daily dosage
veterinary applications	transmucosal administration	sweetener
controlled-	gum	sweeteners

Table 1.A17 **Top 177 text-novel biocides concepts**

Unanimous (included in robustness check)		
thiamethoxam	boscalid	spirodiclofen
azoxystrobin	ethiprole	asulam
clothianidin	methoxyfenozide	noviflumuron
trifloxystrobin	cinosulfuron	thifluzamide
spinosad	penoxsulam	strobilurin
acetamiprid	flonicamid	halofenozide
thiacloprid	triflumuron	oxasulfuron
prothioconazole	neonicotinoid	quinoxyfen
pyraclostrobin	benoxacor	diofenolan
emamectin	isoxaflutole	ethaboxam
emamectin benzoate	tebufenpyrad	trifloxysulfuron
fluquinconazole	sulfosulfuron	gamma-cyhalothrin
dinotefuran	novel active compound	cyazofamid
lufenuron	metaflumizone	dioxygenase
imazamox	dimoxystrobin	fenpyroximate
controlling animal pests	isoxadifen-ethyl	milbemectin
nitenpyram	spiromesifen	cloquintocet
kresoxim-methyl	metosulam	zeta-cypermethrin
mesotrione	pyridaben	bromobutide
ipconazole	teflubenzuron	halosulfuron-methyl
fluoxastrobin	florasulam	thifensulfuron-methyl
sulfentrazone	chlorfluazuron	c1-c4-alkoxy
hexaflumuron	cyclosulfamuron	mefenoxam
chlorfenapyr	imazapic	chlorantraniliprole
cloquintocet-mexyl	protoporphyrinogen	pyroquilon
flumioxazin	protoporphyrinogen oxidase	fluxofenim
tebufenozide	fenclorim	fenhexamid
indoxacarb	orysastrobin	tritosulfuron
famoxadone	penthiopyrad	oxabetrinil
c1-c4-alkyl	spirotetramat	mepronil
mefenpyr-diethyl	flutolanil	tricyclazole
picoxystrobin	isoxaben	thenylchlor
novaluron	carfentrazone-ethyl	acibenzolar-s-methyl
pymetrozine	propaquizafop	aminopyralid
flumetsulam	foramsulfuron	flubendiamide
spinetoram	simeconazole	flufenacet
boxh	pyridalyl	etoxazole
ethoxysulfuron	fenamidone	metominostrobin
diafenthiuron	pyrifenox	isoprothiolane
spiroxamine	tau-fluvalinate	iodosulfuron
triazamate	iprovalicarb	moxidectin
daimuron	fosamine	fosthiazate
iminoctadine	oxadiargyl	diflufenzopyr
fenoxaprop-p	furametpyr	c1-c4-haloalkyl
carfentrazone	doramectin	macrocyclic
phytopathogenic harmful fungi	flufenerim	cyometrinil
phytopathogenic harmful	probenazole	nithiazine
fluopyram	trinexapac-ethyl	bixafen
pyridin-3-yl	diclosulam	isotianil

Table 1.A17 (cont.)

Unanimous (included in robustness check)		
chromafenozide	bifenazate	saflufenacil
cyhalofop-butyl	mandipropamid	fluopicolide
pyributicarb	cyprosulfamide	flupyrsulfuron
kinoprene	cyflufenamid	metalaxyl-m
triazoxide	mepanipyrim	pyriprole
nanoparticles	metrafenone	benthiavalicarb
clodinafop	proquinazid	cyantraniliprole
		tolfenpyrad
Consensus (excluded in robustness check)		
extenders and/or surfactants	preventively	
ch3	r14	
plant essential	fully halogenated	

Table 1.A18 Top 213 text-novel fertilizer concepts

Unanimous (included in robustness check)		
selenium	inorganic substrate component	biomass particles
itaconic	inorganic substrate	organic drying
itaconic moieties	cell component	compound drying
itaconic acid	corn steep	organic compound drying
itaconic anhydride	corn steep liquor	compound drying agent
compost tea	fertigation	biotic
canola	bactericide	co2
canola oil	maleic moieties	vermicompost
particle domain	bioorganic	recurring polymeric
mean particle domain	inorganically augmented bioorganic fertilizer	polymeric subunits
water-dispersible particle	bioorganic fertilizer	recurring polymeric subunits
particle dispersion	inorganically augmented bioorganic	sulfate nitrate
polymer-containing composition	inorganically augmented	ammonium sulfate nitrate
soil amendment compositions	animal manures	wood ash
chlorine dioxide	hydrolyzed animal	plant nutrient content
wetting agents	animal hair	mycorrhizal fungi
phosphite	hydrolyzed animal hair	seed meal
ferrate	urea-formaldehyde polymer	soy meal
sodium ferrate	vinylic polymer	triple super phosphate
calcium ferrate	vinylic	dried residue
potassium ferrate	vinylic polymers	industrial molasses
decompose potassium	polycarboxylated polymer	pharmaceutical fermentation
potassium minerals	polycarboxylated	threonine
decompose potassium minerals	municipal biosolids	ellipsoideus
decompose potassium compounds	biochar	delbrueckii
partial salt	meat meal	saccharomyces delbrueckii
copolymer salt	cerevisiae	green waste
block copolymer	saccharomyces cerevisiae	toxins
yeast cell	saccharomyces cerevisiae hansen	heat source
yeast cells	hansen	abiotic
carbon-skeleton energy	cerevisiae hansen	dissolved materials
carbon-skeleton	calcium hypochlorite	phosphorus minerals

(*continued*)

Table 1.A18 (cont.)

Unanimous (included in robustness check)		
carbon skeleton energy	adenosine	decompose phosphorus minerals
skeleton energy	adenosine triphosphate	decompose phosphorus
complex carbon	triphosphate	biostimulant
carbon compounds	atp	radical polymerization
complex carbon compounds	nh4	free radical polymerization
binder component	ester groups	swine manure
water-soluble binder	environmentally friendly	bio
substrate component	biomass feedstock	dissolved oxygen
metal silicate	saccharomyces uvarum beijer	saccharomyces ludwigii
electrical conductivity	uvarum	saccharomyces willianus
heat-dried biosolids	uvarum beijer	willianus
heat-dried	saccharomyces uvarum	saccharomyces rosei
lower alcohol	beijer	rosei
pva	mellis	rouxii
bactericidal	saccharomyces mellis	saccharomyces rouxii
neodymium	saccharomyces microellipsoides	saccharomyces sake
bifenthrin	microellipsoides	sake
c1-c4 alcohols	oviformis	exiguus
electromagnetic field	saccharomyces oviformis	saccharomyces exiguus
decompose phosphorous	saccharomyces fermentati	carlsbergensis
decompose phosphorous compounds	fermentati	saccharomyces carlsbergensis
aluminum phosphate	saccharomyces logos	chevalieri
organic alcohols	logos	saccharomyces chevalieri
sylvinite	ludwigii	saccharomyces sp.
Consensus (excluded in robustness check)		
tea	energy component	bimodal vinylic
mean particle	skeleton energy component	bimodal vinylic polymer
particle domain size	decompose complex carbon	ch2
polymer-containing	decompose complex	overproduce
biological fertilizer composition	convert complex	overproduce growth
domain size ranges	convert complex carbon	paste-like
mean particle size	binder component present	paste-like material
amendment compositions	steep liquor	dust control
enhancing soil	hemp	drying agent selected
soil fertility	fertilizer marketplace	quick drying
salt form	agricultural fertilizer marketplace	drying properties
partial salt form	agricultural crop	organic drying agent
form granulated particles	commercial agricultural fertilizer	quick drying properties
form granulated	polymer made	integrated system
yeast cell component	polymer composition also	mgso

Table 1.A19 **Top 106 text-novel machinery concepts**

Unanimous (included in robustness check)		
aeration apparatus	operating travel	controller communicatively
axle driving apparatus	operating travel direction	controller communicatively coupled
antenna	wheel configured	air cart
robotic arm	operative control	perimeter wall
flexible cutterbar assembly	rolling basket	upright axes spaced
modular disc cutterbar	teatcup liner	residue spreader
modular disc	foot platform	crop residue spreader
cutterbar assembly flexes	position based	rotary milking
fore-and-aft draper	receiving data	pin configured
flexible draper	meter roller	aeration tine
trimmer head assembly	robotic attacher	aeration pockets
nontransitory computer	axle driving unit	energy storage device
computer readable	agricultural row unit	wireless communication
computer readable medium	forward working direction	crop throughput
nontransitory computer readable	zero radius turning	residue chopper
computer program product	approximate zero	aeration tines
product tank	grain cart	tool coupled
gps receiver	ecu	imaging device
seed metering system	pump mounting surface	inductor box
location-determining receiver	rotary cutting deck	output device
location-determining	computer-readable	distribution lines
gnss receiver	location information	horizontal cutter disks
gnss	motor mounting surface	generally horizontal cutter
		positions spaced transversely
Consensus (excluded in robustness check)		
controller configured	system based	module configured
actuator configured	controllably operable	plate configured
apparatus configured	control unit configured	belt configured
dairy livestock	vehicle position	sensor arrangement
arm configured	agricultural working machine	processor configured
cutterbar assembly attached	valve configured	manner selected
assembly flexes	motor configured	conveyor configured
sickle assembly supported	characteristic data	units configured
controller operable	controller receiving	adjustment mechanism configured
opening configured	chamber configured	controller controlling
harvesting header operable	headland	crop inputs
readable medium	executable	evaluate

Table 1.A20 **Top 118 text-novel plants concepts**

Unanimous (included in robustness check)		
transgene	modified carbohydrate metabolism	genetic material
transgene encoding	acid metabolism	glyphosate
transgene encodes	phenoxy proprionic acid	glufosinate
locus conversion	phenoxy	sulfonylurea
single locus	phenoxy proprionic	transgenic
single locus conversion	proprionic	benzonitrile
backcross conversion	proprionic acid	triazine
backcross	nucleic acid	backcrossing
backcross progeny	nucleic	tissue cultures
progeny plants	altered fatty	bacillus
selected progeny	altered	bacillus thuringiensis
trait selected	altered phosphorus	thuringiensis
selected progeny plants	phosphorus	bacillus thuringiensis endotoxin
herbicide selected	altered carbohydrates	endotoxin
selected backcross	carbohydrates	thuringiensis endotoxin
produce selected	altered fatty acids	pest resistance
selected backcross progeny	fatty acids	dicamba
backcross progeny plants	altered antioxidants	herbicide resistant
higher backcross	antioxidants	imidazolinone
higher backcross progeny	altered essential amino	transgenes
transformation	amino	insect resistant
f1 progeny	amino acids	fungal
insect resistance	essential amino	waxy starch
plant derived	essential amino acids	pistil
soybean hulls	modified protein	root tip
modified fatty acid	protein concentrate	bacterial
modified fatty	protein isolate	viral disease
fatty acid metabolism	herbicide tolerance	hypocotyl
metabolism	abiotic stress	introgressed
carbohydrate metabolism	abiotic	traits introgressed
carbohydrate	abiotic stress tolerance	
modified carbohydrate	herbicide resistance	
Consensus (excluded in robustness check)		
transgene confers	produce backcross progeny	isolate
encoding	locus confers	subsequent generation
transgene conferring	single locus confers	environmental conditions
conversion	plant product	site-specific recombination
locus	commodity plant	recombination
locus conversion confers	commodity plant product	waxy
desired trait	hulls	tip
produce backcross	concentrate	corn variety

Table 1.A21 Top 122 text-novel research tools concepts

Unanimous (included in robustness checks)		
clustal alignment method	hairpin rna	abiotic stress tolerance
clustal method	amplicon	seed-preferred
novel nucleotide	clongasc	hemp
single nucleotide	antibody compositions	digestibility
sequence identity based	nitrogen use efficiency	pufas
identity based	increased biomass	biofuel
gene silencing	increased seed yield	colloid
transcribable polynucleotide	increased oil	agr
transcribable polynucleotide molecule	switchgrass	schizochytrium
isolated polynucleotides	agrobacterium-mediated	transcription factors
chimeric gene results	mediated transformation	lyophilization
pesticidal polypeptide	agrobacterium-mediated transformation	poaceae
polyunsaturated fatty acids	olive	sirna
oilseed plant	isolated polypeptides	salinity
plant biomass	diacylglycerol	epa
nucleic acid segments	diacylglycerol acyltransferase	dalapon
eicosapentaenoic acid	mirna	fescue
eicosapentaenoic	salix	thraustochytrium
docosahexaenoic acid	salix species	pathogen-inducible promoter
docosahexaenoic	crucifers	dehalogenase
acid metabolism	heterologous nucleotide sequences	hppd
acid segments	molecular markers	castor bean
fatty acid metabolism	stress-related protein	coconut palm
wild type variety	carbohydrate metabolism	snp
rnai	fluorescent protein	silage
turfgrass	green fluorescent	starch branching
double-stranded rna	green fluorescent protein	frt
rna interference	vicia species	cosmetics
Consensus (excluded in robustness checks)		
clustal	polynucleotide operably	nitrogen use
clustal v method	isolated polynucleotides encoding	polypeptides encoded
clustal v	coding nucleic	food product
alignment method	coding nucleic acid	primer pair
clustal alignment	acid molecules encoding	stress-related
pairwise alignment	acid molecule operably	gene involved
one regulatory sequence	type variety	full complement
provides recombinant expression	corresponding wild	element operably linked
silencing	full-length complement	agronomic interest
polynucleotide selected	representative seed	recombination sites
isolated polynucleotide selected	encodes seq	orientation relative
one polynucleotide	encodes seq id	increasing resistance
polynucleotide operably linked	use efficiency	

References

Akcigit, Ufuk, Douglas Hanley, and Nicolas Serrano-Velarde. 2021. "Back to Basics: Basic Research Spillovers, Innovation Policy, and Growth." *Review of Economic Studies* 88 (1): 1–43. https://doi.org/10.1093/restud/rdaa061.
Alston, J. M. 2002. "Spillovers." *Australian Journal of Agricultural and Resource Economics* 46 (3): 315–46.
Balsmeier, B., M. Assaf, T. Chesebro, G. Fierro, K. Johnson, S. Johnson, G. C. Li, S. Lück, D. O'Reagan, B. Yeh, and G. Zang. 2018. "Machine Learning and Natural Language Processing on the Patent Corpus: Data, Tools, and New Measures." *Journal of Economics & Management Strategy* 27 (3): 535–53.
Bloom, Nicholas, Mark Schankerman, and John Van Reenen. 2013. "Identifying Technology Spillovers and Product Market Rivalry." *Econometrica* 81 (4): 1347–93.
Chen, Lixin. 2017. "Do Patent Citations Indicate Knowledge Linkage? The Evidence from Text Similarities between Patents and Their Citations." *Journal of Informetrics* 11:63–79.
Clancy, M. S., and G. Moschini. 2017. "Intellectual Property Rights and the Ascent of Proprietary Innovation in Agriculture." *Annual Review of Resource Economics* 9:53–74.
Clancy, Matthew, and Sneeringer, Stacy. 2018. "Eating the Seed Corn? The Impact of Generic Drug Entry on Innovation in Animal Health." Paper presented at the Agricultural and Applied Economics Association annual meeting, August 5–7, 2018, Washington, DC. doi/10.22004/ag.econ.274379.
Evenson, R. E. 1989. "Spillover Benefits of Agricultural Research: Evidence from U.S. Experience." *American Journal of Agricultural Economics* 71 (2): 447–52.
Fuglie, K. O., and P. W. Heisey. 2007. *Economic Returns to Public Agricultural Research*. Economic Brief No. 10, US Department of Agriculture, Economic Research Service.
Gardner, B. L. 2002. *American Agriculture in the Twentieth Century: How It Flourished and What It Cost*. Cambridge, MA: Harvard University Press.
Greenstone, M., R. Hornbeck, and E. Moretti. 2010. "Identifying Agglomeration Spillovers: Evidence from Winners and Losers of Large Plant Openings." *Journal of Political Economy* 118 (3): 536–98.
Griliches, Z. 1957. "Hybrid Corn: An Exploration in the Economics of Technological Change." *Econometrica* 25 (4): 501–22.
———. 1958. "Research Costs and Social Returns: Hybrid Corn and Related Innovations." *Journal of Political Economy* 66 (5): 419.
———. 1992. "The Search for R&D Spillovers." Supplement, *Scandinavian Journal of Economics* 94:29–47.
Huffman, W. E., and R. E. Evenson. 2006. *Science for Agriculture: A Long-Term Perspective*. 2nd ed. Ames, IA: Blackwell.
Jaffe, A. B., and G. de Rassenfosse. 2017. "Patent Citation Data in Social Science Research: Overview and Best Practices." *Journal of the Association for Information Science and Technology* 68 (6): 1360–74.
Jaffe, Adam B., Manuel Trajtenberg, and Michael S. Fogarty. 2000. "The Meaning of Patent Citations: Report on the NBER/Case-Western Reserve Survey of Patentees." NBER Working Paper No. 7631. Cambridge, MA: National Bureau of Economic Research.
Khanna, J., W. E. Huffman, and T. Sandler. 1994. "Agricultural Research Expen-

ditures in the United States: A Public Goods Perspective." *Review of Economics and Statistics* 76 (2): 267–77.
Lampe, Ryan. 2012. "Strategic Citation." *Review of Economic Studies* 94 (1): 320–33.
Latimer, R., and D. Paarlberg. 1965. "Geographic Distribution of Research Costs and Benefits." *Journal of Farm Economics* 47 (2): 234–41.
Li, Danielle, Pierre Azoulay, and Bhaven N. Sampat. 2017. "The Applied Value of Public Investments in Biomedical Research." *Science* 356 (6333): 78–81.
Marx, M., and A. Fuegi. 2019. "Reliance on Science in Patenting." Working paper, Boston University, Boston, MA, February 20, 2019. Available at SSRN.
Moser, Petra, Joerg Ohmstedt, and Paul W. Rhode. 2018. "Patent Citations—an Analysis of Quality Differences and Citing Practices in Hybrid Corn." *Management Science* 64 (4): 1926–40.
Nagaoka, S., K. Motohashi, and A. Goto. 2010. "Patent Statistics as an Innovation Indicator." In *Handbook of the Economics of Innovation*, edited by B. H. Hall and N. Rosenberg, 1083–1127. Amsterdam: North Holland Publishing.
Packalen, M., and J. Bhattacharya. 2015. "New Ideas in Invention." NBER Working Paper No. 20922. Cambridge, MA: National Bureau of Economic Research.
Roach, Michael, and Wesley M. Cohen. 2013. "Lens or Prism? Patent Citations as a Measure of Knowledge Flows from Public Research." *Management Science* 59 (2): 504–25.
Schultz, T. W. 1956. "Reflections on Agricultural Production, Output and Supply." *Journal of Farm Economics* 38 (3): 748–62.
Sneeringer, Stacy, and Matt Clancy. 2020. "Incentivizing New Veterinary Pharmaceutical Products to Combat Antibiotic Resistance." *Applied Economic Perspectives and Policy* 42 (4): 653–73.
United States Bureau of Labor Statistics. 2007. "Technical Information about the BLS Multifactor Productivity Measures." September 26, 2007. https://www.bls.gov/mfp/special_requests/mfptablehis.xlsx.
United States Department of Agriculture (USDA), Economic Research Service (ERS). 2019. "Agricultural Research Funding in the Public and Private Sectors." Last updated September 30, 2019. https://www.ers.usda.gov/data-products/agricultural-research-funding-in-the-public-and-private-sectors/.
———. 2020. "Agricultural Productivity in the U.S." Last updated November 17, 2020. https://www.ers.usda.gov/data-products/agricultural-productivity-in-the-us/.
United States Patent and Trademark Office (USPTO) Patent Technology Monitoring Team. 2019. *U.S. Colleges and Universities—Utility Patent Grants, Calendar Years 1969–2012*. https://www.uspto.gov/web/offices/ac/ido/oeip/taf/univ/org_gr/all_univ_ag.htm.
Wang, C., Y. Xia, and S. Buccola. 2009. "Public Investment and Industry Incentives in Life-Science Research." *American Journal of Agricultural Economics* 91 (2): 374–88.

Comment Alberto Galasso

Economists use the terms *knowledge spillovers* and *research spillovers* to indicate the positive effects that the research and development (R&D) investments of one firm may have on other firms. The idea that research investments generate positive externalities, and thus increase productivity growth and subsequent innovation of other firms, is one of the primary justifications for government R&D support policies.

How to identify and measure research spillovers is one of the classic research questions in the field of economics of innovation. For many decades, researchers struggled to find a way to measure empirically these spillovers. Krugman (1991, 53) wrote that knowledge spillovers "are invisible; they leave no paper trail by which they may be measured and tracked, and there is nothing to prevent the theorist from assuming anything about them that she likes."

Empirical scholars responded to Krugman, documenting and leveraging a variety of paper trails in the forms of citations in patents and scientific publications. This generated a vibrant, large, and growing literature.[1] Clancy, Heisey, Ji, and Moschini contribute to this stream of research, providing a thoughtful examination of knowledge spillovers from nonagricultural technologies into agricultural innovation.

The chapter employs three different empirical measures of knowledge spillovers. The first measure exploits citations made by patent documents. Consider a patent protecting an agricultural technology that cites many prior patents that are not classified by the patent office as agricultural technologies. In this case, the citation pattern suggests that knowledge spillovers from outside agriculture were important for the development of the innovation. The chapter builds on this idea and also leverages the richness of the patent data to measure the specialization of the firm owning the cited patent. As more agricultural patents cite firms that are not specialized in agriculture, support for the idea that there are important knowledge spillovers from other industries grows stronger.

The second measure of spillovers presented in the chapter relies on patent citations to scientific publications. The intuition behind this measure is that citations from agricultural patents to nonagricultural academic jour-

Alberto Galasso is a professor of strategic management at the University of Toronto, where he holds the Rotman Chair in Life Sciences Commercialization, and a research associate of the National Bureau of Economic Research.

For acknowledgments, sources of research support, and disclosure of the author's material financial relationships, if any, please see https://www.nber.org/books-and-chapters/economics-research-and-innovation-agriculture/comment-roots-agricultural-innovation-patent-evidence-knowledge-spillovers-galasso.

1. See Bloom, Schankerman, and Van Reenen (2013) for a recent contribution and a description of the various empirical approaches developed in the literature.

nals reveal that academic research in other scientific domains has significant knowledge spillover into agriculture.

The final measure is based on a text-analysis algorithm that identifies the appearance of new "textual concepts" (i.e., text strings) on agricultural patents. With this approach, the presence of knowledge spillovers is revealed by textual concepts that are new in agricultural patents but are not novel in other technology fields.

The empirical analysis in the chapter suggests that knowledge spillovers from outside agriculture are a statistically significant and economically important driver of agricultural innovation. A large fraction of these spillovers appear to be derived from biology and chemistry, two research fields that are technologically close to agriculture.

The large spillovers documented by Clancy, Heisey, Ji, and Moschini have important implications for our understanding of how shocks propagate in the economy through industry linkages. There is a growing literature examining how supply-and-demand shocks that originate in one industry may percolate through vertical chains or disseminate to other industries (Barrot and Sauvagnat 2016; Galasso and Luo 2018). The results described in the chapter show strong research linkages between agriculture and other technology areas, which suggest that agricultural innovation may be exposed to shocks in these research domains.

To develop some policy implications, it is important to understand the channels through which knowledge is transmitted to (and from) agricultural research. Numerous studies in the economics of innovation literature implicitly assume that knowledge flows are not tradable and that the empirically measured research spillovers only capture unintended external effects. While this may be an appropriate assumption in some contexts, it may not be valid in many technology sectors. In the presence of well-functioning markets for technology, knowledge may be transmitted across firms through patent licensing contracts. Moreover, firms may leverage their intellectual property assets to facilitate knowledge exchanges with some fields but not others. As explained in a recent study by Arque-Castells and Spulber (2019), to understand the role played by the market for technology, it is essential to assess the wedge between the social and private rates of return of R&D. Combining data on out-of-field citations with data on patent licensing, reassignment, and litigation may help us understand the extent to which knowledge flows are internalized.

The innovation literature has stressed the importance of general purpose technologies (GPTs). These are inventions that have potential applications across a wide number of sectors (Bresnahan and Trajtenberg 1995). Examples of GPTs include the steam engine, the electric motor, microprocessors, and more recently, artificial intelligence. GPTs have been shown to be powerful sources of growth in sectors that can develop complementary technologies. The literature has documented substantial heterogeneity across sectors in this respect. These differences are typically linked to market structures and appropriability conditions. In light of these findings, an interpretation of the results described by Clancy, Heisey, Ji, and Moschini is that the agricultural sector has been very effective at exploiting GPTs originating in other sectors. In principle, the high rate of GPT adoption by agricultural innovators may have enhanced the innovation incentives in the GPT itself (Cockburn, Henderson, and Stern 2019).

The estimates in the chapter show that the percentage of prior-art citations that accrue to patents not classified as agricultural patents is very large in some agricultural subsectors. For example, about 90 percent of patents cited by animal health patents are not classified by the US Patent and Trademark Office (USPTO) as agricultural patents. This is a striking result. One important thing to notice, though, is that interpreting the magnitude of citation-based measures of spillovers is challenging. This is because it is not clear what the appropriate benchmark should be. As a reasonable first step, Clancy, Heisey, Ji, and Moschini examine whether the fraction of citations made to nonagricultural patents is above or below 50 percent. Technology areas in which more than half of the cited references belong to other fields are highlighted as fields receiving large external knowledge spillovers. A more general analysis of this issue may require benchmarking the propensity of agricultural patents to cite out-of-the-field patents with similar propensity measures in other technological areas.

From a conceptual perspective, one also has to consider the possibility that the magnitude of spillover effects may be determined by the relative size of a technology field. This may be particularly important when two research areas are technologically very close but differ in size. Consider the following example in which there are two technology fields, field A and field B. In field A, there are 10 patents, and in field B, there are 90 patents. Now assume that each of these 100 patents randomly cites one of the other 99 patents. In this case, if citations are independent and identically distributed, one would observe many more patents in field A citing patents in field B than patents in field B citing patents in field A. At the same time, the high propensity of field A patents to cite out-of-the-field patents is not really revealing that each invention in field A builds disproportionally from field B. It is simply reflecting the fact that A is a small field, with fewer knowledge inputs to draw from, and heavily connected to the larger field B.

In conclusion, Clancy, Heisey, Ji, and Moschini make a convincing case that ideas that originate outside of agriculture have important effects on agricultural research, perhaps a role as important as R&D investments within agriculture. They also provide a variety of different and powerful empirical measures to capture knowledge flows into agriculture. Future research should focus on further understanding the drivers and implications of these important findings.

References

Arque-Castells, P., and D. Spulber. 2019. "Measuring the Private and Social Returns to R&D: Unintended Spillovers versus Technology Markets." Northwestern Law & Econ Research working paper, Chicago, IL.

Barrot, J., and J. Sauvagnat. 2016. "Input Specificity and the Propagation of Idiosyncratic Shocks in Production Networks." *Quarterly Journal of Economics* 131:1543–92.

Bloom, N., M. Schankerman, and J. Van Reenen. 2013. "Identifying Technology Spillovers and Product Market Rivalry." *Econometrica* 81:1347–93.

Bresnahan, T., and M. Trajtenberg. 1995. "General Purpose Technologies 'Engines of Growth'?" *Journal of Econometrics* 65:83–108.

Cockburn, I., R. Henderson, and S. Stern. 2019. "The Impact of Artificial Intelligence on Innovation." In *The Economics of Artificial Intelligence: An Agenda*, edited by A. Agrawal, J. Gans, and A. Goldfarb, 115–46. Chicago: University of Chicago Press.

Galasso, A., and H. Luo. 2018. "When Does Product Liability Risk Chill Innovation? Evidence from Medical Implants." NBER Working Paper No. 25068. Cambridge, MA: National Bureau of Economic Research.

Krugman, Paul R. 1991. *Geography and Trade*. Cambridge, MA: MIT Press.

2

Quantifying Heterogeneous Returns to Genetic Selection
Evidence from Wisconsin Dairies

Jared Hutchins, Brent Hueth, and Guilherme Rosa

2.1 Introduction

Biological innovation is an important driver of productivity growth in the agricultural sector (Olmstead and Rhode 2008). This is especially so in the dairy sector, where milk yield has grown 3–4 percent per year during the past century; 50 percent of this growth is typically attributed to improvement in livestock genetics (Pryce and Veerkamp 2001). However, growth can be overattributed to genetic improvement when models ignore the fact that dairy farmers select genetics based on their farm's returns to a given type of genetics. This is because in the dairy sector, the vast majority of "experimentation" undertaken to identify high-performing genetics takes place in *nonexperimental conditions*. Starting in 1908, the US Department of Agriculture (USDA) initiated a program, in partnership with land grant universities and local associations of dairy farmers, to measure and record animal-level performance (CDCB 2017). This partnership, which eventually came to be known as the Dairy Herd Improvement (DHI) program, continues to this day. Data from commercial dairy herds that participate in the DHI

Jared Hutchins is an assistant professor in the Department of Agricultural and Consumer Economics at the University of Illinois at Urbana-Champaign.

Brent Hueth is a branch chief of structure, technology and productivity at the US Department of Agriculture, Economic Research Service.

Guilherme Rosa is a professor in the Department of Animal and Dairy Sciences at the University of Wisconsin-Madison.

The findings and conclusions in this publication are those of the authors and should not be construed to represent any official USDA or US Government determination or policy. This research was supported in part by the US Department of Agriculture, Economic Research Service.

For acknowledgments, sources of research support, and disclosure of the authors' material financial relationships, if any, please see https://www.nber.org/books-and-chapters/economics-research-and-innovation-agriculture/quantifying-heterogeneous-returns-genetic-selection-evidence-wisconsin-dairies.

program (roughly half of all US dairy herds) are used to estimate the "genetic merit" of individual male dairy cattle (sires). Specifically, USDA scientists use regression models to identify the impact of a sire on the performance of its offspring. Because of these *nonexperimental* conditions, these public estimates of sire productivity can contain both the quality of the genetics and the skill of the farmer in selecting them to the right environment. Without understanding the contribution of farmer management and selection to productivity, productivity growth in the dairy sector may be mistakenly attributed to the work of breeders instead of to both breeders and farmers.

Our work decomposes total productivity change on Wisconsin dairy farms due to genetics into separate effects for genetic improvement and endogenous selection. Using data from a large sample of Wisconsin dairy farms and national-level data on dairy sire rankings, we develop and estimate a model that accounts for selection behavior in the animal's production function. We find that selection accounts for as much as 75 percent of the total productivity improvement in fat and protein yield in dairy cows. Our results provide evidence for positive assortative matching, whereby farmers who adopt above-average yield genetics also perform better than average for their chosen genetics. Further, we find that management behavior accounts for a significant portion of within-herd cow-level heterogeneity in genetic choice, suggesting that, contrary to previous evidence, farmers manage at the animal level and not solely at the herd level; this implies that previous regression models controlling only for herd-level variation do not adequately control for selection bias in the production function. Overall, our results indicate that a large portion of productivity growth in dairy farming can be explained by farmers' ability to identify and select genetics that are well suited to their production environment—not solely the quality of the genetics they choose.

Using on-farm data to calculate the genetic merit of different sires, thus far the engine of genetic progress in the dairy industry, runs the risk of entangling the management savvy of dairy farmers with the quality of the genetics. Artificial insemination (AI) technologies, widely adopted beginning in the 1960s, expanded opportunities to track and identify the performance of sires. Modern AI technologies permit a single sire to produce hundreds of thousands of offspring, and each female offspring (birthed on the farm of a DHI program participant) contributes new sire evaluation data, improving estimates of genetic merit. Genetic merit in turn strongly influences the market price AI companies receive for their genetic material. However, the very thing that makes dairy unique also makes isolating the contribution of genetics tricky. Research by economists on technology adoption has shown that productivity gains can be overattributed to technology when there is "positive assortative matching," meaning the very ones that benefit the most from technology are the ones who will adopt it. Hybrid seeds, for example, were adopted into their most productive environments in the United States (Griliches 1957) as well as in Kenya (Suri 2011). Similarly, technologies such as fertilizer are often applied by farmers who have the highest returns from applying it (Foltz, Aldana, and Laris 2014).

Our work modifies the canonical modeling framework used by quantitative geneticists for determining genetic merit by accounting for the selection behavior of dairy farmers. Our research bridges two scientific domains, both indebted to the seminal work of Sewall Wright, that diverged early in the 20th century in their approaches to explaining the contribution of genetics in the farm production function. Using a control function approach, we find that the average returns to the adoption of high-yield genetics on dairy farms are as much as 75 percent lower after accounting for confounding factors. We find evidence of positive assortative matching at the cow level and the herd level, suggesting that dairy farmers manage their herds at the level of individual cows. Our work makes a novel contribution to the literature on technology adoption by investigating a level of detail in selection behavior that is previously unexplored in the agricultural sector.

We start with the idea that farmers select genetics based on their ex ante returns to the technology, which will cause them to "match" sire genetics to cows in a specific way. We assume further that some of the factors affecting match quality are observed by farmers but not by researchers. Using the framework of the correlated random coefficient (CRC) model, we explore the effect of this selection behavior on returns to production traits in dairy cattle and test whether correcting for selection behavior affects the estimation of the average effect. We use random variation in country-wide genetic evaluations as an instrument to identify the effect of choosing dairy sires with high predicted transmitting ability (PTA) indices for fat and protein yield in the cow's production function, and we use the residual from the first stage to identify the heterogeneity in the effect. We find that the average return from increasing the index one unit, which is a one-unit increase in pounds of fat or protein production, is .6 and .4 for fat and protein without controlling for selection behavior. These estimates drop to .15 and .18 after controlling for selection behavior, which means that as much 75 percent of the return to high-yield genetics is explained by this matching behavior. Finally, we find that the heterogeneity in returns is over both farms and animals; this implies the high productivity gains are being driven by animal-level matching and not just farm-level matching. This changes the narrative of farm productivity in the dairy industry quite drastically: instead of the triumph of animal breeders and scientists alone, it is growth accomplished by a partnership between farmers and breeders.

2.2 Related Literature

Though seldom pointed out, the estimation of breeding values in quantitative genetics and production function estimation in economics share a common history. The roots of both can be traced back to Sewall Wright, who pioneered work in population genetics, paving the way for the field of modern quantitative genetics (Gianola and Rosa 2015). Wright also conceptualized the notion of "path analysis," which later evolved to be known as "instrumental variables," and applied it to supply and demand systems as well as simultaneous equation models of commodity prices (Wright 1928; Wright 1925). From this common origin, economics and quantitative genetics diverged in point of focus with respect to estimating the following equation:

$$(1) \qquad y = Z\mu + X\beta + \varepsilon,$$

where y is the output (e.g., butterfat production of a dairy cow or yield of a strain of maize), Z is an incidence matrix of genetic technologies or types, X is a matrix of "environment" covariates unrelated to genetics, and ε is the unexplained component of y.

In quantitative genetics, the parameters β are modeled as fixed, but the parameter μ is treated as the outcome of a genetic process and thus considered a random variable with a covariance matrix mapping all the relationships among genotypes. Nowhere has this genetic model developed quite the importance it has in animal breeding as a result of the work of C. R. Henderson (1953, 1973). Prior to Henderson's work, there was no widely used method for attributing the performance of different livestock to its parents. The Henderson mixed model (HMM), still used in the US national DHI program, models breeding values as draws from the random variable μ (Henderson 1975).

The HMM has become integral to the dairy genetics industry because estimates of μ for each sire, $\hat{\mu}$, strongly influence market prices for dairy genetics. PTA, which is $\hat{\mu}/2$, is roughly interpreted as the value that a sire has for a particular trait y, which is predicted to be "transmitted" to the offspring (Van Vleck 1987). The national DHI program produces PTA values for a wide variety of traits including milk yield, fat yield, fertility, longevity, and "conformance" (elements of body structure such as udder size and height). Once published, these values influence adoption decisions, which then result in new data that feed back into the DHI program as raw data to create breeding values for new sires and to update estimates for breeding values of existing sires. Building from Sewall Wright, the HMM has become an important source of genetic progress for the dairy industry.

The field of economics developed in parallel to Henderson's work but focused on a different set of estimation issues with respect to equation (1). In particular, the production function literature in economics has centered on the assumptions needed to identify estimate μ. If the adoption of certain genetics is associated with unobserved components of y, this means $\mathrm{Cov}(Z,\varepsilon) \neq 0$, and standard regression approaches yield biased estimates of μ.

This bias was more generally referred to by Mundlak (1961) as "management bias," defined as the presence of unobserved management decisions (or conditions of the decision environment) that influence input choice by farmers (genetic selection in this context). Griliches (1957) specifically suggested in the case of hybrid corn that genetic technology was historically adopted into the environment where it was the most profitable. Solutions to this problem have evolved from the simple fixed effects approach of Mundlak (1961) to invoking the wisdom of Sewall Wright and using exogenous variation to identify structural parameters of production functions (Griliches and Mairesse 1995).

More recently, labor economists have developed new frameworks for thinking about this identification issue. The Roy model (Roy 1951) posited that occupation decisions, much like technology adoption decisions, are not chosen randomly; instead, they are generated from behavior that takes into consideration ex ante idiosyncratic returns that are difficult to measure. This implies that measuring the returns to some decision on an outcome, such as the effect of the adoption of technology on firm output, is subject to a "selection bias" that must be dealt with in something like equation (1) (Heckman and Vytlacil 1998). A similar logic can be applied to the choice of genetics, since farmers likely observe or know something ex ante, unobserved to researchers, and affecting relative returns across relevant genetic profiles. Suri (2011) formalized the link between labor economics and production function estimation by using the Roy model to study selection bias in technology adoption. Her study found that farmers in Kenya adopted hybrid maize if their personal unobserved return was high, suggesting "positive assortative matching." This has in turn helped spur a growing literature on quantifying the heterogeneous returns to agricultural technology adoption in other contexts (Foltz, Aldana, and Laris 2014; Michler et al. 2019; Zeitlin et al. 2010).

Our analysis circles back to an empirical question that has been studied for nearly 100 years: How do we evaluate the performance of animal genetics from observational data? We unite two divergent fields of study, economics and quantitative genetics, by bringing the insights and theory of economic analysis to the wealth of data on dairy animal performance and genetics and its associated modeling approaches. Returning to the basic structure of the HMM, we focus on estimating the effects of the genetic indices, PTAs, for production traits in dairy cattle and whether estimation suffers from selection bias. If PTAs are affected by selection behavior, this indicates that part of dairy farm productivity usually attributed to genetic progress should also be attributed to farmer skill at matching genetics to their environments.

In the next section, we provide a theoretical framework for thinking about heterogeneous returns to dairy genetics and how their effect on productivity can be investigated using Wooldridge's (2015) CRC model.

2.3 Theory and Methodology

To begin, consider the case where choosing genetics is equivalent to choosing to increase or decrease a single trait by purchasing a sire with a particular PTA value. Every sire can be described as a vector of PTA index values for various traits, and in this case, we can think of genetic selection as the choice of a vector of index values. In reality, choosing a sire is a discrete decision, as the farmer faces some choice set of sires from various AI companies. We assume in what follows that the space of PTAs is "dense enough" that a farmer can choose any level of the trait they want from their choice set independently of other traits. We further assume this decision is only based on the trait itself and not on the sire's identity or on the AI company that is offering it for sale.

Studying adoption via a continuous variable is preferable to the discrete approach in this case because it is not known which sires are in the farmer's choice set. There are more than 10,000 unique sires in our data, and many more are actually available to farmers. Future analysis of the space of sires may be able to find a reduced dimensionality representation suitable for discrete choice analysis. As a first attempt, we study only the adoption of the traits via the PTA index to be able to apply a wider range of econometric tools.

2.3.1 Theoretical Framing

The following simple model demonstrates the role of farm- and animal-level heterogeneity in estimating the average returns to genetic investments via a continuous index. Unlike other input decisions, the decision to invest in genetics by choosing a certain sire happens three years before the animal starts producing. Assume that there is only one trait, z, which the producer has to choose three years before the cow begins production to maximize ex ante expected return:

$$\max_z \bar{\pi}(z, x, v) - wz,$$

where x and v are observable and unobservable management at the farm level, $\bar{\pi}$ is expected lifetime profit, and w is the price of purchasing one more unit of a trait.

In this model, the choice of z is only affected by *farm-level* heterogeneity, v and x. This is the level at which heterogeneity is usually analyzed based on the notion that management decisions operate at the level of an entire farm (Mundlak 1961; Suri 2011). What the above does not consider is that the characteristics of the mate—that is, the animal that is bred with the sire—should also affect returns to z.

Call these unobserved animal-level characteristics u. We can modify the above model only slightly to show why these characteristics are important.

Instead of z affecting $\overline{\pi}$ directly, it operates indirectly through a transmission function $f(z,u)$, which takes the traits of the father, or "sire" (z), and the traits of the mother, or "dam" (u), and maps to a new trait value, z':

$$\max_z \overline{\pi}(z',x,v) - wz$$

$$s.t.\ z' = f(z,u).$$

Now the optimal choice of z depends on the current period price that is to be paid versus the expected increase in profit weighted by how well the trait transmits. Adding this transmission function implies that unobserved heterogeneity affecting the adoption of z operates at the farm and animal levels. This is an important distinction and departure from the assumptions of both economic and animal science models of the returns to adoption. Economic models of the effect of technology adoption consider heterogeneity at the firm or farm level due to the assumption that confounding variation is from management behavior affecting all plots and animals. Animal science models refer to confounding variation at the animal level as "preferential treatment" and generally only control for farm-level effects because the literature does not find substantive evidence of animal-level decision-making that would bias evaluations (Graham, Smith, and Gibson 1991; Tierney and Schaeffer 1994).

Despite the lack of attention in the literature, animal-level heterogeneity can play its own part in biasing evaluations. If the manager observes components of u that are unobserved in data, then he or she may invest z with animals where the return is highest. Our next step is to investigate how the existence of unobserved u will affect our empirical evaluations of the returns to z in a production function.

2.3.2 Empirical Model

In data, we observe farms j and cows i during time period t. The PTA value of the sire chosen for an animal, z_{ij}, is time invariant. Using the above framework, we assume that there are animal-level (u_{ij}) and farm-level (v_j) "match quality" components that affect return to adoption of z_{ij}. Assume further that the total return is linearly separable, such that the total return is . Define this payoff as relative to some average expected return $\overline{\mu}$ so that u_{ij} and v_j are the farm's known deviation from this average, $\tilde{\mu}_{ij} = u_{ij} + v_j$. The returns to z_{ij} for a given animal i and farm j are thus $\mu_{ij} = \overline{\mu} + \tilde{\mu}_{ij}$.

Heterogeneity in the production function manifests in the coefficients for z_{ij}:

$$y_{ijt} = \mu_{ij} z_{ij} + \beta X_{ijt} + \varepsilon_{ijt}.$$

Assuming a constant slope to identify $\overline{\mu}$ has the following effect on the equation:

$$\begin{aligned} y_{ijt} &= (\bar{\mu} + \tilde{\mu}_{ij})z_{ij} + \beta X_{ijt} + \varepsilon_{ijt} \\ &= \bar{\mu}z_{ij} + \beta X_{ijt} + (\tilde{\mu}_{ij}z_{ij} + \varepsilon_{ijt}) \\ &= \bar{\mu}z_{ij} + \beta X_{ijt} + \xi_{ijt}. \end{aligned}$$

Because $\tilde{\mu}_{ij}$ is unobserved match quality, then ordinary least squares will not identify an unbiased $\bar{\mu}$. It is biased because the variable return to a trait $\tilde{\mu}_{ij}z_{ij}$ is in the error term so that $E(z_{ij}\xi_{ijt}) \neq 0$. Instrumental variables will also not identify $\bar{\mu}$ because anything correlated with z must be correlated with ξ (Cornelissen et al. 2016).

Our identification strategy in this chapter uses instead the control function method and its specific approach to random coefficients, the CRC model. With this approach, we approximate input demand with a linear function of observed covariates plus an excluded variable and use the residual from the approximation to proxy for match quality in the production function. Wooldridge (2015) spells out two main conditions for the control function method to identify $\bar{\mu}$ and uncover heterogeneity in the effect of a trait. Defining η_{ij} as the residual term from a linear approximation of trait demand for z_{ij}, the two conditions for the CRC model are as follows:

$$\text{A1: } E(\varepsilon_{ijt} \mid \eta_{ij}) = \rho\eta_{ij}$$

$$\text{A2: } E(\tilde{\mu}_{ij} \mid \eta_{ij}) = \psi\eta_{ij}$$

These are both strict assumptions about how informative the residual η_{ij} is in capturing bias and match quality. A1 is a standard assumption for control function methods and says that selection bias takes a particular form: the conditional expectation of unobserved components of output is linear in η_{ij}. A2 says that the heterogeneous slope coefficient defined across cows must be proportional to the input demand residual η_{ij}. The unobserved components to technology adoption must include $\tilde{\mu}_{ij}$ if the manager considers their ex ante returns when choosing z_{ij}, but this assumption restricts their relationship to be proportional. Using A2, we can use η_{ij} interacted with z_{ij} to proxy for an estimate of $\tilde{\mu}_{ij}$. Either of these assumptions can be relaxed to be nonlinear, but explicit functional forms must be given so that we know how to include η in the production function. In our analysis, we maintain the linear forms.

We also need an exogenous shifter of z_{ij} that is uncorrelated with ε_{ijt}. Our instrument is the difference between the sire's PTA at the time it was chosen and its PTA value at its next evaluation four months after the adoption date (t'): $\Delta z_{ij} = z_{ij}^{t'} - z_{ij}$.[1] PTAs for every sire are updated by the Council on Dairy Cattle Breeding (CDCB) every four months using herd testing data from

1. Special thanks to our discussant, Paul Scott, for this suggestion.

around the country. The change in PTA from one evaluation to the next Δz_{ij} is linearly related to z_{ij} and so is a relevant predictor, but the size of the deviation has to do with the performance of the sire's daughters all across the country. This deviation is likely unrelated to unobserved production ε_{ijt} because it is based on the performance of other offspring of the sire before ε_{ijt} is ever realized. It is also unlikely that the updates happening right after the use of the genetics will somehow influence future management of that offspring; if this were the case, the PTA value of that sire at the time the offspring *ij* starts producing would be the more actionable information rather than the intermediate updates. For these reasons, we believe Δz_{ij} satisfies the exclusion restriction needed for an instrument. Our approach shares similarities to the control function approaches of Levinsohn and Petrin (2003) and Olley and Pakes (1996), which also use dynamic input lags as an exogenous source of variation to identify production function parameters.

Using A1 and A2 and the instrument Δz_{ij}, we can adjust the production function for the bias resulting from heterogeneous returns. Defining $\hat{\eta}_{ijt}$ as the estimated residual from the first-stage input demand, we now write our empirical model as two equations:

$$z_{ij} = \alpha_0 + \gamma \Delta z_{ij} + \beta_0 X_{ijt} + \eta_{ijt}$$

$$y_{ijt} = \alpha_1 + \bar{\mu} z_{ij} + \rho \hat{\eta}_{ijt} + \psi \hat{\eta}_{ijt} \times z_{ij} + \beta_1 X_{ijt} + \varepsilon_{ijt}$$

- y_{ijt}: dairy cow performance for butterfat/protein in a given lactation
- z_{ij}: value of PTA butterfat/protein of the sire chosen at the time of adoption
- Δz_{ij}: deviation in trait value in the next updated evaluation
- X_{ijt}: time-varying management decisions affecting y (a full list can be found in appendix A)
- η_{ijt}: input demand residual

Our main research question has to do with the parameters $\bar{\mu}$, ρ, and ψ. The hypothesis of "perfect transmission" of a trait is that $\bar{\mu} = 1$, so a one-unit increase in PTA causes a one-unit increase in the offspring's performance (Kearney et al. 2004). How much this parameter differs from one before and after our bias correction indicates whether unobserved management decisions affect the average return to a trait in our sample. If ρ is statistically different than zero, this rejects the null hypothesis that the models with and without the correction are equivalent (Wooldridge 2015).

Finally, ψ indicates the relationship between match quality and returns to z. If $\psi > 0$, then cows matched with a sire that has higher-than-expected PTA will also have a higher marginal return to PTA in their production function. This is consistent with the "positive assortative matching" story, which is that farmers adopt traits that work particularly well on their farms.

2.3.3 Heterogeneity Distribution

An output of the above model is an estimate of μ_{ij}, $\hat{\mu}_{ij} = \bar{\mu} + \psi\hat{\eta}_{ijt}$.[2] Assuming our theoretical framework from before, this estimate contains both μ_{ij} and v_j. The farm-specific component v_j has been the focus of most studies in economics and is controlled for in animal science using fixed effects (termed "contemporary groups" in the animal science literature). However, we may also be interested in how much of the distribution in returns is driven by the animal-specific component u_{ij}. If there are heterogeneous returns at the animal level, then this means that the returns to the adoption of genetics are diverse even within a given farm environment. It also implies that sire evaluation models using farm fixed effects do not completely control for confounding factors and that there is evidence of managers matching specific genetics to specific animals, which could bias estimates of the return to genetics.

After estimating the parameters $\bar{\mu}$, ρ, and ψ, we estimate the distribution $\hat{\mu}_{ij}$ using the CRC model with three different specifications: no fixed effects, herd fixed effects, and herd-by-time fixed effects. The first specification estimates a distribution that contains both u_{ij} and v_j, and the second nets out v_j. The third specification mimics the fixed effects strategy of many genetic evaluation models, which use a herd fixed effect interacted with the time of the observation to soak up dynamic management decisions affecting the returns to genetics.

2.4 Data

As described above, the market for dairy sires makes heavy use of CDCB evaluations, which are calculated from DHI data. In addition to milk yield and somatic cell count, the DHI program tracks the number of times per day each cow is milked (usually two, sometimes three), their calving and birth dates, and their "lactation number" (the number of lactation cycles a cow has been through at a given point in time). Unfortunately, no other management decisions are observed. Our current data set covers DHI herds served by one dairy records processing center in the state of Wisconsin from June 2011 to January 2015, which is representative of about 40 percent of Wisconsin dairy herds. At the lactation level, there about 1 million lactation records for approximately 277,000 dairy cows on 1,500 dairy farms.

Because of the lack of management decisions observed in DHI data, the HMM includes in X a number of fixed effects to attempt to control for the confounding impact of management on genetics. In this model, we control

2. Note that this is in contrast to HMM, which would assume a normal distribution for such an effect and center it at zero. We gain flexibility with the distribution of the coefficient only because we specify exactly what determines the distribution, which is the unobserved variation in input demand.

Table 2.1 **Records description**

Herds	1,459
Sires	7,628
Dairy cows	277,695
Number of lactations	424,910
Lactation records	1,065,308

for lactation length, lactation number, and proportion of lactation milked three times a day in our specification. This is also a "cohort" effect, which is an interaction between herd and test month, which is a herd-specific time trend. There are also biological factors such as birth year, calving month, and breed that are included as fixed effects. In addition to these controls, we also include prices such as the milk price, the ration cost, and the price of replacement heifers. In our main specification, we use only herd fixed effects but also estimate the model with herd-by-time effects as a robustness check when looking at relevant distributions.

Every cow that shows up in DHI data has an ID that connects back to a sire and associated evaluation available from the CDCB. The CDCB updates evaluations three times a year, and these PTA values are the ones that will appear to the farmer when choosing genetics. Sire evaluations are publicly available on the CDCB's website and are reported by AI companies when selling sires. These evaluations are updated four times a year. Using the sire IDs in our data set, we recovered all available records of these sires throughout time and matched them to cow records. Thus for each cow in our data, we know the PTA value of its parent sire at the time the choice of sire was made. The "time they were chosen" is calculated as 10 months prior to the cow's birth date to account for the gestation period of a dairy cow. Our data set contains more than 7,000 unique dairy sires matched to our 1 million lactation records.

We use the folowing prive covariates in our model. For ouput and input price, we use "income over feed cost," a relationship between milk price and ration cost determined by the 2006 farm bill. We also include the price of 16 percent dairy ration as an additional proxy for feed cost. Finally, we include the cost of replacement, which we calculate as the beef price per pound times 1,400, the typical weight of a dairy cow, minus the cost of a replacement heifer.

In addition to issues discussed thus far, the analysis of dairy cow lactation records is complicated by survival bias for cows on their second lactation onward. Managers may remove cows from their herd if they do not meet some threshold of production during the first lactation. This selection issue is discussed in detail in Henderson (1975), and often lactation records of cows past lactation one are not used in sire evaluations for this reason. Keeping this in mind, we implement practices commonly followed in dairy

Table 2.2 **Covariate description**

	Mean	SD
Continuous variables		
PTA fat	28.79	27.12
PTA protein	21.46	20.48
Proportion milked 3×	0.58	0.49
Lactation length	310.44	23.48
Herd size	157.35	232.99
Binary variables (%)		
Lactation number		
1st	45.73	
2nd	28.71	
3rd	15.35	
4th	7.34	
5th	2.87	

science literature when analyzing lactation data. We do not consider cows that are lactation six or higher (about .1 percent of the data), and we analyze "primiparous" (first lactation) cows separately from "multiparous" cows. Primiparous cows should not be subject to survival bias, while multiparous cows are a subset of the first group that was not culled. It should be the case that multiparous cows are more subject to the management bias we discuss, and we analyze this group separately to see how our bias correction works differently in this subset. If bias is severe in multiparous cows, this suggests an interaction between the behavior affecting genetic selection and the behavior affecting culling decisions.

Using our matched data, we graph the kernel densities of PTA values for butterfat and protein chosen in this sample with dotted lines indicating their average. Recall that the HMM used to produce PTA measures fixes the distribution to be normal with a mean of zero for the relevant population. The densities are not centered at zero and are not symmetric; both densities have very long left tails, which shift averages to the left.

This does not give any indication of at what level the selection is occurring, however. For example, do farmers simply choose the same PTA value for all their animals in a month? At what level is there variation in the chosen traits? A quick calculation of within-group sum of squares can shed light on how variable each selected trait is within a given herd versus between herds. For example, if a farmer simply chooses the same trait value for all his or her cows, then the sum of squares within herds should be zero. The farmer may choose the same trait in all time periods or, within a certain month, choose the same value for all cows. The difference between herd-month and herd essentially approximates the importance of time-variant factors, such as prices and other economic factors. We calculate the ratio between the within

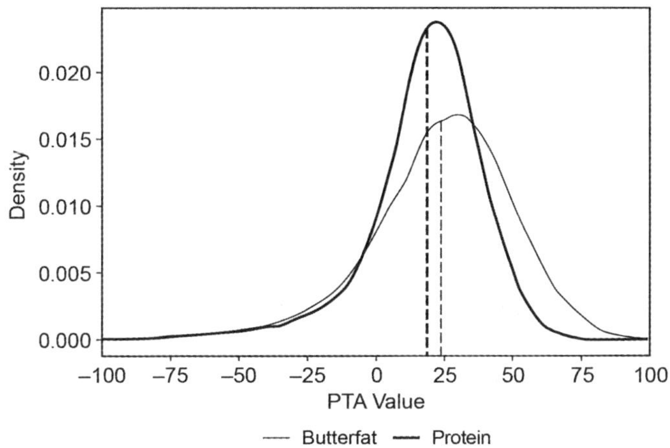

Fig. 2.1 Distributions of PTAs in the data

Table 2.3 Proportion of SST explained

	PTA fat	PTA protein
Herd	0.954920	0.955163
Herd by time	0.811062	0.797823

sum of squares for these two groups (herd-adoption month and herd) and the total sum of squares.[3]

The proportion of the sum of squares total explained by within-herd variation in choices of both traits is quite large: about 80 percent for herd-adoption month and around 95 percent for herd. This is evidence that the largest amount of variation in trait choices is within a herd and not between herds, or a large variation at the cow level. This is not consistent with a model where heterogeneity in selection behavior is driven at the herd level. Given this indicative evidence, we proceed to our empirical model to explore the impact of this heterogeneity in selection on the average return to high-yield genetics.

2.5 Results

We study the traits protein and butterfat, which are the components of milk that are most important to profitability for dairy farmers in Wisconsin. The PTA index is in units of pounds of fat and protein and represents the

3. Calculated as $\Sigma_{g=1}^{G}\Sigma_{i=1}^{I}(y_{ig} - \bar{y}_g)^2 / \Sigma_{g=1}^{G}\Sigma_{i=1}^{I}(y_{ig} - \bar{y})^2$, where G is either herd groups or herd adoption month group.

expected increase in yield of a dairy cow using the given sire relative to a base sire (whose PTA is zero). Specifically, they are predictions of a statistical model, the HMM, which are interpreted as the increase in fat or protein for the specific sire that is chosen. Since the outcome y_{ijt} is measured in the same units as PTA, if $\bar{\mu} = 1$, then increases in sire ability correspond one-to-one with increases in the offspring's ability.

For each trait, we estimate several different specifications to examine how the coefficient on z changes with different corrections. We estimate ordinary least squares (OLS), OLS with herd fixed effects, two-stage least squares, two-stage least squares with herd fixed effects, and correlated random coefficients (i.e., including $\hat{\eta}_{ijt}$ and $\hat{\eta}_{ijt} \times z_{ij}$ as regressors in a fixed effects regression). If heterogeneity is only at the herd level, then according to Wooldridge (2005), herd fixed effects alone should identify the average treatment effect. Including the control function terms ($\hat{\eta}_{ijt}$ and $\hat{\eta}_{ijt} \times z_{ij}$) in the fixed effects model identifies the heterogeneity within herds specifically, so the difference between these specifications provides evidence regarding the importance of cow heterogeneity in determining the average effect.

Finally, noting that the marginal benefit of a trait is given by

$$\hat{\mu}_{ijt} = \bar{\mu} + \hat{\psi}\hat{\eta}_{ijt},$$

we can graph the resulting distribution to examine the variability of returns across the entire sample. In each specification, we analyze three samples based on the lactation year: all lactations, first lactations, and later lactations. The first-lactation cows are studied separately because they are not subject to survival bias, as later-lactation cows possibly are. Estimates of the first stage of the model from which the input demand residual is calculated are presented in appendix B. Standard errors are calculated clustered at the herd level and cluster bootstrapped for the CRC model.

2.5.1 Fat

For both OLS and fixed effects, the average return to increasing the butterfat of a sire is positive and different than 0. It is around 0.6, meaning a one-unit increase in a pound of PTA causes a 0.6-pound increase in offspring. The correction, however, attenuates the effect toward 0 by a large amount. When using instrumental variables and a constant coefficient on z_{ij}, the coefficient is near 0, implying that the correction takes away most of the productivity gain that would otherwise be (mis)attributed to the choice of PTA. The average effect identified in the CRC model is higher, about 0.14, with a positive and significant ρ (meaning there was significant selection bias in the OLS specification). The CRC specification also tells us that cows with a higher than predicted amount of the trait have a higher marginal return to the trait—that is, $\psi > 0$. At all levels, we reject the hypothesis that $\rho = \psi = 0$, which indicates that the model with the correction is statistically different than the model without it. This suggests that the instrument was necessary

Table 2.4 **All lactations**

	OLS (1)	FE (2)	IV (3)	IV + FE (4)	CRC + FE (5)
PTA fat	0.604***	0.544***	0.0325**	0.0355***	0.149***
	(0.0321)	(0.0134)	(0.00723)	(0.00434)	(0.0083)
$\hat{\eta}$					0.563***
					(0.0236)
$\hat{\eta} \times$ PTA fat					0.0066***
N	1,065,308	1,065,308	1,065,308	1,065,308	1,065,308
Adj. R^2	0.351	0.562	0.345	0.557	0.564

Standard errors in parentheses. * $p < 0.10$, ** $p < 0.05$, *** $p < 0.01$.

Table 2.5 **Across lactation cows**

	All (1)	First lactation (2)	Later lactation (3)
PTA fat	0.14933***	0.10593***	0.19799***
	(0.00827)	(0.00829)	(0.01303)
$\hat{\eta}$	0.56271***	0.53914***	0.58808***
	(0.02364)	(0.02447)	(0.03091)
$\hat{\eta} \times$ PTA fat	0.00662***	0.00602***	0.00724***
	(0.00037)	(0.00042)	(0.00044)
N	1,065,308	511,446	553,859
Adj. R^2	0.564	0.537	0.514

Standard errors in parentheses. * $p < 0.10$, ** $p < 0.05$, *** $p < 0.01$.

to correct the estimates for endogeneity. These results imply that the average effect of investing in high-fat genetics is reduced by 75 percent after taking into account confounding factors. More than half of the return to PTA fat is explained by unobserved confounding variables.

Table 2.5 shows a similar pattern to the whole sample and further indicates large differences in first- and later-lactation cows for average returns. There is strong evidence of selection behavior that affects the returns to genetics, as the average return for first-lactation cows is half that of later-lactation cows. If this difference is generated by culling, it indicates that farmers cull cows in their first lactation that have low marginal return to the high-yield genetics.

2.5.2 Protein

Similar to fat, the average returns to protein are much lower when accounting for confounding factors. Using simple OLS, the return to protein is 0.427 and indistinguishable from 0 when using instrumental variables. Using the CRC model, the effect is different than 0 but is less than half of the OLS

Table 2.6 **All lactations**

	OLS (1)	FE (2)	IV (3)	IV + FE (4)	CRC + FE (5)
PTA protein	0.427***	0.358***	0.00883	0.0165***	0.21724 ***
	(0.0343)	(0.0109)	(0.00603)	(0.00381)	(0.011)
$\hat{\eta}_{ijt}$					0.25252***
					(0.0102)
$\hat{\eta}_{ijt}$ × PTA protein					0.00983***
N	1,065,308	1,065,308	1,065,308	1,065,308	1,065,308
Adj. R^2	0.451	0.669	0.448	0.667	0.671

Standard errors in parentheses. * $p < 0.10$, ** $p < 0.05$, *** $p < 0.01$.

Table 2.7 **Across lactation cows**

	All (1)	First lactation (2)	Later lactation (3)
PTA protein	0.21724***	0.19005***	0.23834***
	(0.011)	(0.01104)	(0.0134)
$\hat{\eta}$	0.25252***	0.2472***	0.26652***
	(0.02248)	(0.02271)	(0.02875)
$\hat{\eta}$ × PTA protein	0.00983***	0.00997***	0.00972***
	(0.00041)	(0.00042)	(0.0005)
N	1,065,308	511,446	553,859
Adj. R^2	0.671	0.632	0.622

Standard errors in parentheses. * $p < 0.10$, ** $p < 0.05$, *** $p < 0.01$.

coefficient when controlling for confounding factors: the estimate changes from 0.43 to 0.18. The direction of ψ suggests positive assortative matching for the adoption of high-protein genetics, just as for high-fat genetics. When looking at different lactations, there is less evidence of culling based on returns to protein. There are slightly lower returns for first-lactation cows than later-lactation cows, but it is a much smaller difference compared to the differences for fat.

2.5.3 Distributions

Here we estimate the resulting distributions from the CRC specification, $\hat{\mu}_{ijt} = \bar{\mu} + \hat{\psi}\hat{\eta}_{ijt}$. Figure 2.2 shows the distribution of returns across all lactations for fat and protein. Figure 2.3 shows the distributions of both traits across different lactations and across different levels of fixed effects. In addition to using herd fixed effects, we also include a combination of three fixed effects for herd, test month, and calving month. The product of these three indicator variables is typically referred to as a "contemporary group effect"

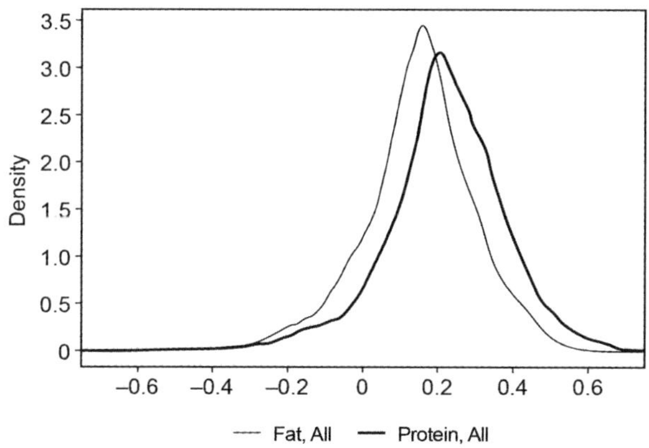

Fig. 2.2 Distributions of returns

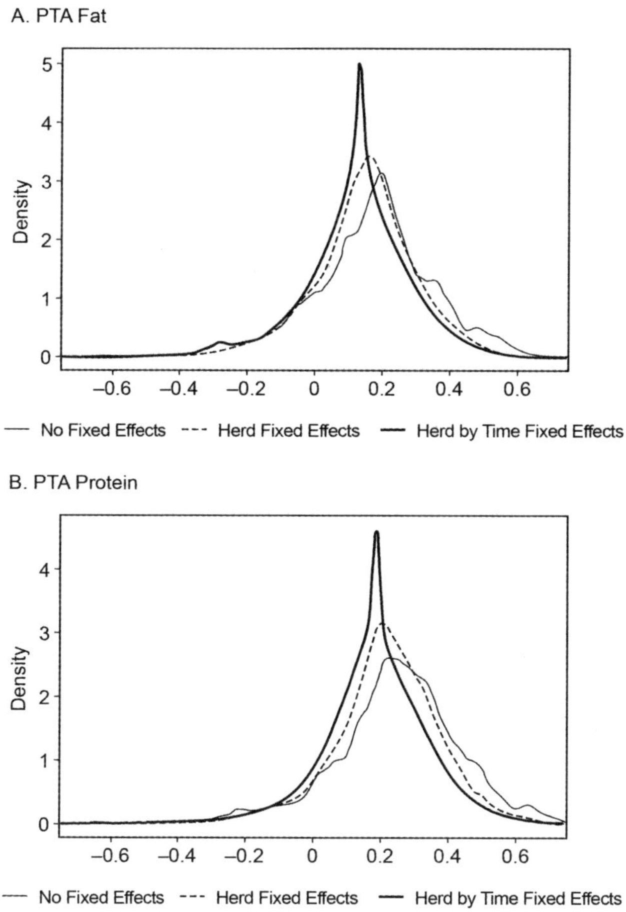

Fig. 2.3 Fat and protein distributions

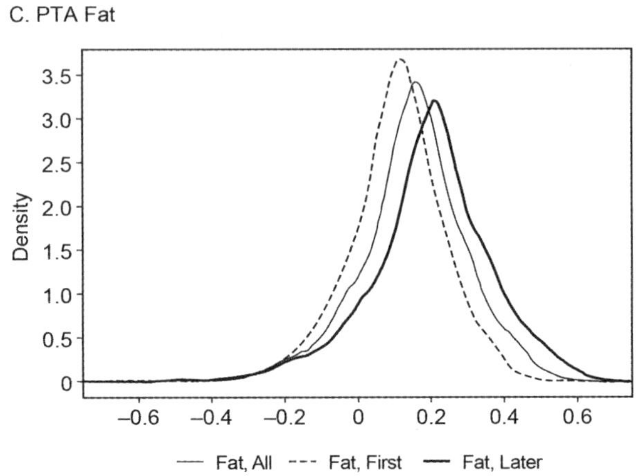

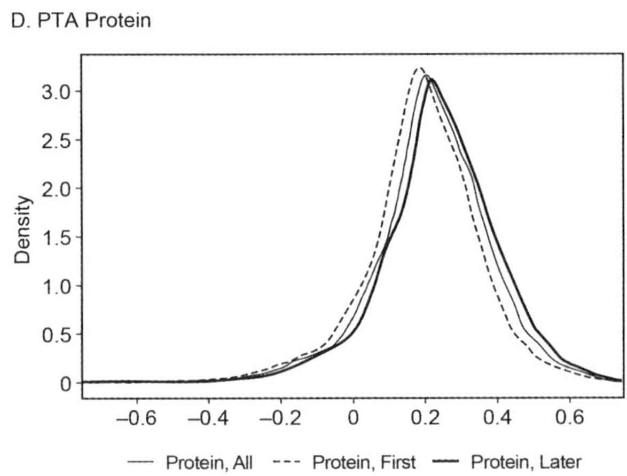

Fig. 2.3 (cont.)

in a genetic evaluation model. Table 2.8 shows the means and standard deviations of all of the distributions.

While different in their average effect, both traits have about the same standard deviation. For fat, a part of the distribution actually has a negative coefficient for adoption. For that farm or animal, the returns from adopting the technology may in fact be negative because of a combination of management environment and unobserved animal-level factors. In this case, however, it is hard to justify why increasing a trait in one parent would actually decrease that same trait in the offspring. The reason for this may be that high-fat or high-protein genetics are correlated with another trait that may negatively affect milk production in certain environments. We made

Table 2.8 **Distributions of marginal returns**

	Fat		Protein	
	Mean	Std. dev	Mean	Std. dev
All lactations	0.150	0.151	0.218	0.164
First lactations	0.106	0.133	0.191	0.161
Later lactations	0.199	0.168	0.239	0.166
No fixed effects	0.170	0.185	0.249	0.197
Herd fixed effects	0.150	0.151	0.218	0.164
Herd-by-time fixed effects	0.132	0.146	0.187	0.158

the assumption that traits could be chosen independent of one another, but in reality, traits have genetic correlations. For example, high milk yield and health are negatively correlated, so increasing production traits could negatively impact health, causing a decrease in phenotypic yield. A more advanced model of adoption would need to find a way to model their adoption together and explicitly include these correlations as part of the choice problem.

Figure 2.3 shows the differences in returns across lactations and across different fixed effects specifications. The distributions across lactations support the results in tables 2.5 and 2.7. The goal of using different fixed effects specifications was to look at the effects of parsing out herd-level unobserved factors v_j versus animal-level unobserved factors u_{ij} from the distribution $\hat{\mu}_{ij}$. Without any fixed effects, the effect sizes are 0.17 and 0.25 for fat and protein, respectively. After netting out time-invariant herd effects, the effect sizes drop to 0.15 and 0.218. Finally, when netting out *all* herd-level variation using herd-by-time effects, the effect sizes drop to 0.132 and 0.187. The resulting distributions are entirely generated by variation in input demand at the animal level, meaning dairy farm managers observe animal-level returns that they use to choose genetics. These results imply that, on average, these returns make up 77 percent and 75 percent of the variable returns to investment in high-yield genetics. A sizable portion of the selection process, therefore, appears to happen at the level of animals rather than at the level of farms.

2.6 Discussion and Conclusion

We examine the effect of economic selection behavior on the returns to the adoption of genetic technology for dairy farms in Wisconsin. Previous literature has attributed a large amount of productivity growth on dairy farms to improvements in genetics without considering the possibility that traits are selected into environments because of unobserved (to the researcher) farm- or animal-specific returns to a given trait. Using the theoretical framework of the Roy model, we argue that farm- and animal-level heterogeneity may bias estimates in the returns to genetic traits such as butterfat and protein through the selection process. We use herd testing association data on dairy cows and the evaluations of their sires at the time they were chosen to estimate a CRC model. Our approach permits us to examine the effects of correcting for bias when estimating the impact of genetic improvement on productivity, the effect of selection behavior as a source of heterogeneity in returns to traits, and the relative importance of cow-level heterogeneity versus farm-level heterogeneity.

We find that correcting for selection bias lowers the estimated contribution of genetic improvement to productivity differences across cows by an average of 50 percent. We estimate average returns to adopting genetics with 1 more pound of fat or protein to be about 0.6 pounds and 0.4 pounds before the correction and 0.15 pounds and 0.2 pounds after correction for selection bias. Our model also indicates positive assortative matching, meaning farms with the highest return to adopting a given set of traits are the ones that adopt, giving credence to the upward bias in the coefficients. We also find that first-lactation cows have the lowest average return to high-yield genetics, indicating that farmers tend to cull cows with lower-than-expected ex post marginal return to the traits. Analyzing the distributions of returns, we find that up to 75 percent of the heterogeneity in returns generated from the input demand residual were at the animal level instead of the farm level. This implies that the factors confounding the returns to genetics are also at the animal level; cows with different trait investments are managed differently in ways that are not controlled for using farm-level fixed effects.

This study has several limitations that should be addressed in future work. First, we model trait adoption as though each trait could be chosen independent of other traits. This may not be a reasonable assumption given the extent to which traits are correlated with one another. Accounting for this possibility would require using a system of equations with cross-equation restrictions limiting trait-selection possibilities to those that are feasible given relevant empirical context. Such a model might be able to explain why there are negative returns to high-yield genetics for some animals and farms. Second, we treat trait investment as continuous even though farms choose sires discretely. This assumes that the trait values are dense enough to treat the variable as continuous, whereas the adoption decision is discrete over a choice set of individual bulls. One way to model this as a discrete problem while also taking into account correlations between traits would be to use a lower dimensionality representation of sires determined from the data. Unsupervised machine learning methods such as K-means clustering could be used to characterize an implied grouping of sires that have certain traits in common. The problem of choosing traits would then become one of choosing a basket of traits represented by a certain grouping of bulls. The matching decision itself also needs further elaboration. In future work, we hope to turn our attention to data on breeding decisions where we have additional data that may permit a more detailed investigation of the selection decision. We also aim to develop a more sophisticated modeling approach that will take into account that farmers select a portfolio of traits rather than choosing one at a time.

Despite these shortcomings, our results point to new possibilities for studying technology adoption and suggest the need for a reinterpretation of and further research on the expansive literature that examines the contribution of genetic progress to productivity growth in the dairy sector. For the economics field tackling technology adoption, animal-level heterogeneity is important and should not be overlooked. Appendix B contains the first-stage results of the model, which show that cows with higher trait investment are milked more frequently and survive to more lactations. If such behavior happens at the animal level, it is important to take this into account when thinking of sources of farm productivity that farmers may act on in the context of the Roy model. Previous studies of farm productivity usually identify "unobserved" returns at the farm level, and for this reason, many papers studying dairy farms or animal operations sum production to the herd level. This assumption also suggests a reevaluation of extension programming developed to advise farmers about herd-level management. Our work shows that a large amount of the heterogeneity in returns is driven by animal-level variation, meaning there are ample opportunities to increase productivity by emphasizing management on this level. Agricultural data are becoming more granular, and there is no doubt there will be increasing opportunities for economics research to take selection of genetics by farmers into account. We consider only animal agriculture, where every animal must be bred, but the approach we develop here may also be used at some scale in crop agriculture.

Overall, we find that selection behavior biases estimates of the effect that genetic improvement alone has on productivity growth. An important component of productivity change depends on farmers choosing genetics that work particularly well in conditions that are idiosyncratic to their individual farming operations. This changes the narrative regarding the source of farm productivity in the dairy industry from one where science alone is the source of gains from new technology to one where growth is the result of complementary inputs provided by farmers and scientists. Indeed, the success of the dairy industry thus far depends on collaboration among farmers and scientists via institutions often taken for granted, such as the DHI program, land grant universities, and a variety of industry collaborators (represented collectively by the CDCB). The interplay among these organizations and the remarkable record of success (as measured by productivity growth) they have achieved make the dairy industry a unique model of research and innovation in agriculture that merits further analysis and critique in the economics field.

Appendix A

Regression Controls

To select controls for the animal equation, we draw on the animal science literature to inform controls we include in the model.

The vector X_{ijt} contains the following variables:

- Economic Controls
  - cost of 16 percent dairy ration
  - income over feed cost
  - replacement cost (beef price \$/lb × 1,400 − cost of replacement heifer)
  - time trend
- Biological Controls
  - calving month (indicator)
  - test month (indicator)
  - birth year (indicator)
  - lactation number (indicator)
  - Holstein (indicator)
- Management Controls
  - proportion of lactation milked three times a day
  - herd size (deviations from average)
  - lactation length (days in milk of record)

Appendix B

First-Stage Estimates

The first-stage equation for our model uses past variation in a sire's evaluation, which occurs at the national level as a source of exogenous variation:

$$z_{ij} = \alpha_{0j} + \gamma \Delta z_{ij} + \beta_0 X_{ijt} + \eta_{ijt}.$$

While this prediction is time invariant (the selection occurs only once), the residual η_{ijt} will still be time variant because of the term X_{ijt}. Due to the presence of X_{ijt}, the first stage essentially treats the same cow at different points in time as entirely separate cows who happen to have the same values of z_{ij}. This means that when we examine the PTA investment for one cow at two

Table 2.9 **First stage regression**

	PTA fat		PTA protein	
	OLS	FE	OLS	FE
Δz_{ij}	0.499***	0.500***	0.495***	0.496***
	(0.000648)	(0.000613)	(0.000550)	(0.000514)
Lactation no. = 2	1.589***	0.368*	1.793***	0.954***
	(0.265)	(0.207)	(0.193)	(0.157)
Lactation no. = 3	2.220***	−0.195	2.789***	1.132***
	(0.519)	(0.409)	(0.376)	(0.310)
Lactation no. = 4	3.164***	−0.393	3.675***	1.254***
	(0.766)	(0.594)	(0.569)	(0.461)
Lactation no. = 5	3.839***	−0.787	4.718***	1.610***
	(0.999)	(0.786)	(0.746)	(0.605)
Proportion milked 3×	1.528***	−0.310	1.210***	−0.0692
	(0.481)	(0.687)	(0.351)	(0.382)
Herd size	0.0000154	−0.000717	−0.0000608	−0.000657
	(0.000258)	(0.000572)	(0.000198)	(0.000519)
Lactation length (days)	0.00640***	0.00561***	0.00559***	0.00522***
	(0.00108)	(0.000913)	(0.000781)	(0.000678)
Holstein	1.475	3.191***	3.260***	2.707***
	(1.475)	(0.561)	(0.791)	(0.656)
Observations	1,641,022	1,641,022	1,641,022	1,641,022
Adjusted R^2	0.249	0.303	0.281	0.333

Standard errors in parentheses. * $p < 0.10$, ** $p < 0.05$, *** $p < 0.01$.

different lactations, it essentially treats these as two adoption decisions; by deciding to let the animal keep producing, the manager implicitly adopts the genetics again. This is not necessarily problematic, but we must estimate the model on these surviving animals separately to understand how the culling decision interacts with the adoption of genetics.

One implication of this approach is that the first stage will help us understand the trait investments for animals that survive. Table 2.9 shows the results of the first stage and the coefficients on animal-level variables. Both OLS and fixed effects are shown to get a sense of what level of variation is important. For example, both lactation length and lactation number are significant in predicting z, which implies that cows that have a larger trait investment are milked longer and are more likely to not be culled in their first year. Milking the cow three times per day is significant in the OLS specification but not in the fixed effects, implying that farms that choose higher investments in production traits also milk more intensively at the herd level. Holstein cows are also most likely to have the highest investment in production traits, which is to be expected given their comparative advantage in high-volume production.

Differences across production traits are mostly seen in the culling decision. Without herd fixed effects, cows that are kept past the first lactation have higher trait investment for both fat and protein. Once herd fixed effects are used, fewer differences are seen across lactations considering only animal-level variation. For fat, only second-lactation cows have marginally more fat investment than first-lactation cows. For protein, all later-lactation cows have higher investments in protein (on the order of one pound more). One thing that can be learned from these results is that adoption and other management decisions are inextricably linked. Specifically, cows that have a high PTA investment are more likely to be kept, milked longer, and milked more intensively.

References

Cornelissen, T., C. Dustmann, A. Raute, and U. Schonberg. 2016. "From LATE to MTE: Alternative Methods for the Evaluation of Policy Interventions." *Labour Economics* 41:47–60.

Council on Dairy Cattle Breeding (CDCB). 2017. *History of USDA Dairy Evaluations*. Accessed September 24, 2019. https://queries.uscdcb.com/aipl/history/hist_eval.htm.

Foltz, J., U. Aldana, and P. Laris. 2014. "The Sahel's Silent Maize Revolution: Analyzing Maize Productivity in Mali at the Farm Level." In *African Successes*, vol. 4, *Sustainable Growth*, edited by Sebastian Edwards, Simon Johnson, and David N. Weil, 111–36. Chicago: University of Chicago Press.

Gianola, D., and G. J. M. Rosa. 2015. "One Hundred Years of Statistical Developments in Animal Breeding." *Annual Review of Animal Biosciences* 3 (1): 19–56.

Graham, N. J., C. Smith, and J. P. Gibson. 1991. "Investigation of Preferential Treatment for Milk Yield in Canadian Holsteins." *Canadian Journal of Animal Science* 71 (1): 21–27.

Griliches, Z. 1957. "Hybrid Corn: An Exploration in the Economics of Technological Change." *Econometrica, Journal of the Econometric Society* 25 (4): 501–22.

Griliches, Z., and J. Mairesse. 1995. "Production Functions: The Search for Identification." NBER Working Paper No. 5067. Cambridge, MA: National Bureau of Economic Research.

Heckman, J., and E. Vytlacil. 1998. "Instrumental Variables Methods for the Correlated Random Coefficient Model: Estimating the Average Rate of Return to Schooling When the Return Is Correlated with Schooling." *Journal of Human Resources* 33 (4): 974–87.

Henderson, C. R. 1953. "Estimation of Variance and Covariance Components." *Biometrics* 9 (2): 226–52.

———. 1973. "Sire Evaluation and Genetic Trends." *Journal of Animal Science* 1973 (issue symposium): 10–41.

———. 1975. "Best Linear Unbiased Estimation and Prediction under a Selection Model." *Biometrics* 31 (2): 423–47.

Kearney, J. F., M. M. Schutz, P. J. Boettcher, and K. A. Weigel. 2004. "Genotype × Environment Interaction for Grazing Versus Confinement. I. Production Traits." *Journal of Dairy Science* 87 (2): 501–9.

Levinsohn, J., and A. Petrin. 2003. "Estimating Production Functions Using Inputs to Control for Unobservables." *Review of Economic Studies* 70 (2): 317–41.
Michler, J. D., E. Tjernström, S. Verkaart, and K. Mausch. 2019. "Money Matters: The Role of Yields and Profits in Agricultural Technology Adoption." *American Journal of Agricultural Economics* 101 (3): 710–31.
Mundlak, Y. 1961. "Empirical Production Function Free of Management Bias." *Journal of Farm Economics* 43 (1): 44–56.
Olley, G. S., and A. Pakes. 1996. "The Dynamics of Productivity in the Telecommunications Equipment Industry." *Econometrica* 64 (6): 1263.
Olmstead, A. L., and P. W. Rhode. 2008. *Creating Abundance: Biological Innovation and American Agricultural Development*. Cambridge, MA: Cambridge University Press.
Pryce, J., and R. Veerkamp. 2001. "The Incorporation of Fertility Indices in Genetic Improvement Programmes." *BSAP Occasional Publication* 26 (1): 237–49.
Roy, A. D. 1951. "Some Thoughts on the Distribution of Earnings." *Oxford Economic Papers* 3 (2): 135–46.
Suri, T. 2011. "Selection and Comparative Advantage in Technology Adoption." *Econometrica* 79 (1): 159–209.
Tierney, J., and L. Schaeffer. 1994. "Inclusion of Semen Price of the Sire in an Animal Model to Account for Preferential Treatment." *Journal of Dairy Science* 77 (2): 576–82.
Van Vleck, L. 1987. "Contemporary Groups for Genetic Evaluations." *Journal of Dairy Science* 70 (11): 2456–64.
Wooldridge, J. M. 2005. "Fixed-Effects and Related Estimators for Correlated Random Coefficient and Treatment-Effect Panel Data Models." *Review of Economics and Statistics* 87 (2): 385–90.
———. 2015. "Control Function Methods in Applied Econometrics." *Journal of Human Resources* 50 (2): 420–45.
Wright, P. G. 1928. *Tariff on Animal and Vegetable Oils*. New York: Macmillan.
Wright, S. 1925. *Corn and Hog Correlations*. Department Bulletin No. 1300, US Department of Agriculture. https://www.biodiversitylibrary.org/bibliography/108042#/summary.
Zeitlin, A., S. Caria, R. Dzene, P. Jansky, E. Opoku, and F. Teal. 2010. "Heterogeneous Returns and the Persistence of Agricultural Technology Adoption." CSAE Working Paper Series 2010–37, University of Oxford Centre for the Study of African Economies.

3

Yield Performance of Corn under Heat Stress
A Comparison of Hybrid and Open-Pollinated Seeds during a Period of Technological Transformation, 1933–55

Keith Meyers and Paul W. Rhode

3.1 Introduction

The advent of commercially viable hybrid corn seeds in the 1930s preceded a rapid rise in US corn yields over the rest of the 20th century. This technology spread and quickly replaced the once predominant open-pollinated seed varieties grown in the United States. Zvi Griliches's (1957) pathbreaking work used the example of hybrid corn seeds to explain patterns in technological diffusion. Griliches hypothesized that hybrid seeds had a fixed productivity advantage over open-pollinated seeds and increased the potential yield ceiling of corns. Hybrid corn adoption started where (open-pollinated) yields were initially higher, and adoption patterns radiated out from these areas. Other observers including Culver and Hyde (2001) and Sutch (2008, 2011) claim that hybrid corn seeds performed better relative to open-pollinated seeds principally during conditions of drought. Academic research, however, has not determined to what extent hybrid seeds mitigated the effects of drought and heat stress (temperatures generally associated with reductions in corn yields and drought-like conditions).[1] Using uncovered

Keith Meyers is an assistant professor of economics at the Danish Institute for Advanced Study (D-IAS) and in the Department of Business and Economics at the University of Southern Denmark.

Paul W. Rhode is a professor of economics at the University of Michigan and a research associate of the National Bureau of Economic Research.

For acknowledgments, sources of research support, and disclosure of the authors' material financial relationships, if any, please see https://www.nber.org/books-and-chapters/economics-research-and-innovation-agriculture/yield-performance-corn-under-heat-stress-comparison-hybrid-and-open-pollinated-seeds-during-period.

1. For corn, daily temperature averages in excess of 29°C are associated with reductions in corn yields (Schlenker and Roberts 2009; Schauberger et al. 2017). We also use drought measures as reported in the Palmer drought severity index (PDSI) as a measure of weather stress.

archival records, we study how prolonged periods of abnormal temperatures and dryness affected yields for the two types of corn seed. We use both variables denoting drought and deviations in agronomic measures of growing degree days (GDDs).

The work of economic historians entails the development and rediscovery of novel data sources. Such archival resources contain reams of informative records with granular details that are not available in official digitally curated publications. Through archival work, economic historians can answer otherwise unanswerable questions. In our efforts to understand the hybrid diffusion story, we have located, digitized, and organized a treasure trove of unpublished manuscripts reporting on hybrid corn diffusion and performance at a more granular geographic level than is available in US Department of Agriculture (USDA) publications. Using unpublished documents contained in Zvi Griliches's personal manuscripts and field trial data buried in obscure Iowa experimental station reports, we construct a panel of hybrid and open-pollinated corn yields. With these records, we can ascertain whether hybrid seeds exhibited drought tolerance or if Griliches's assumption that hybrid seeds increased the yield potential overall were correct.

Understanding the relative performance of hybrid versus open-pollinated corn seeds during periods of drought informs our understanding of the mechanisms driving the diffusion of the new technology. The Pioneer Hi-Bred Corn Company introduced the first successful commercial hybrid corn seeds in the early 1930s during a period of extreme farm distress, historically low commodity prices, and adverse weather conditions. While hybrid seeds cost two to three times more than their open-pollinated counterparts, they quickly replaced open-pollinated corns. If hybrid corns exhibited drought tolerance, then those traits could explain the rapid diffusion of hybrid seed technology in response to the distress caused by the Dust Bowl droughts of the 1930s (Dowell and Jesness 1939; Crabb 1947).

Past research studying hybrid corn adoption starts with the pathbreaking work of Zvi Griliches (1957, 1958, 1960, 1980). Griliches's analyses posited that the profitability of the new seed technology, as captured by expected yield improvements, drove adoption. Even though hybrid seeds diffused across the Corn Belt and Great Plains during a period of extreme drought, Griliches did not investigate the effect of weather on adoption. In his preferred specification, Griliches assumed that the new hybrids were superior to the existing open-pollinated varieties by a multiple that did not vary significantly over time, across regions, or over weather conditions.

More recent research contests Griliches's account and suggests that drought shocks in 1934 and 1936 accelerated hybrid adoption (Culver

The PDSI captures drought-like conditions over multiple months due to excess temperatures and water deficits. The PDSI is strongly correlated with measures of soil moisture (Dai and NCAR 2019).

and Hyde 2001; Sutch 2008, 2011). Richard Sutch notes that hybrid seeds remained relatively expensive during the 1930s—a period of historically low commodity prices—and that the geographic pattern of hybrid corn diffusion shows a dependence on local weather conditions. Narrative evidence adds further support. To cite one example, a *New York Times* headline read in 1940, "50% of Corn Crop in Hybrid Species . . . Agricultural Marketing Service Lays Its Popularity to Drought Resistance."[2] Indeed, Sutch (2008, 2011) highlights the USDA's role in promoting the adoption of hybrid seed technology and argues that hybrid corn's tolerance to drought conditions made the technology more salient for farmers. The economic stress of the Great Depression and extreme droughts of the 1930s eroded the wealth of farmers. One would expect slower hybrid adoption under such circumstances. Richard Hornbeck (2012) finds that many of the adaptive responses to the Dust Bowl were relatively slow. In comparison, from 1931 onward, US farmers rapidly adopted hybrid corn. This turn toward hybrids may have mitigated some of the adverse effects of the Dust Bowl. Switching to hybrids was costly. Nevertheless, the varieties produced by hybrid breeders promised beneficial qualities, including higher yields, shortened the time to maturity, stronger root systems, thicker stalks, disease resistance, and drought tolerance.

3.2 Factors Driving Hybrid Adoption and the (Potential) Yield Advantage of Hybrid Corn Seeds

The story of hybrid corn has been told many times (Crabb 1947; Fitzgerald 1990; Kloppenburg 1988; Olmstead and Rhode 2008). For economists, the starting point is Griliches (1957). In his seminal article, Griliches analyzed this "invention of a way to invent" and mapped estimated parameters of the diffusion process into economic variables of supply and demand. He viewed the diffusion process as primarily a shift between two equilibria over time rather than as a shift of equilibria. He fit logistic curves to annual diffusion data for states and crop reporting districts, reducing the differences across regions to differences in three parameters: origins, slopes, and ceilings.[3] The origin represented the year (relative to 1940) when diffusion in an area crossed the 10 percent adoption threshold. Griliches related the origin date to the "availability" of hybrid seed—and more specifically, to

2. "50% of Corn Crop in Hybrid Species . . . Agricultural Marketing Service Lays Its Popularity to Drought Resistance," *New York Times*, September 10, 1940. The text noted the hybrid's advantages of both drought resistance and higher yields.

3. The analysis covered 31 (out of 48) states and 132 (out of 249) crop reporting districts (CRDs) in the period up to 1956. The USDA's Agricultural Marketing Service (ASM) made available unpublished data for the CRDs. Griliches restricted his analysis to observations between 0.05 and 0.95 of his estimated ceiling level, K. The ceiling was estimated in an admittedly ad hoc way by picking the K that makes the resulting diffusion curves plotted on logistic graph paper look linear.

supply-side forces, including the profitability of seed producers, the cost of innovation, and the potential market density. He related the slope (or speed of diffusion) and the ceiling levels to demand-side forces, specifically to the profitability to farmers of using the new seed. Griliches found that the estimated speed of adoption was rather uniform but declined as one moved away from the center of the Corn Belt. The origin date and ceiling level also declined with distance from the center.

Griliches (1957) argued the diffusion process could be interpreted in a way that was consistent with rational, long-run, profit-seeking behavior by seed producers and farmers. He made no reference to adverse weather shocks or the drought-resistance qualities of hybrid varieties.[4] According to his preferred specification, hybrids promised a time- and region-invariant yield increase—in the range of 10–15 percent—over existing open-pollinated varieties. He further argued that including the changing advantages of the new seed, the prices of corn output, or the prices of hybrid seed would add "nothing of significance" to the explanation of the diffusion process.[5]

Griliches (1957) tabulated but did not use USDA data on the prices (per bushel) of hybrid and open-pollinated seed by state (box 59).[6] He argued that the hybrid seed prices did not vary significantly across space and could be ignored in his analysis of the rate of diffusion (which was modeled as a transition between two equilibria). His treatment of hybrid seed prices is problematic for several reasons. The leading seed companies, especially in the early periods, possessed some market power to set hybrid prices. The farmer's adoption decision relied not on the hybrid corn price alone but on the hybrid seed price relative to other prices—for example, the price of open-pollinated seed. In figure 3.1, the average price of hybrid seed at the state level is approximately 2 to 3 times greater than the average price of open-pollinated seed. Over the 1937–57 period, the coefficient of variation of the price of hybrid seed across states averaged approximately 10 percent. The coefficient of variations of the ratio of hybrid to open-pollinated seed was substantially higher, averaging 16 percent. Griliches also ignored changes over time. In the late 1930s, hybrid seed cost about 3.5 times as much as open-pollinated seed. By the mid-1950s, the ratio had fallen roughly in half, to about 1.8 times. Griliches tabulated but did not use state-level data on seeding rates (box 59).[7] Again, he argued the cross-state variation was negligible. The coefficient of variation of seeding rates in bushels per acre was around 18.7 percent.

4. Although weather conditions clearly affected the "availability" of seed on the supply side and the drought-resistance qualities of new seed impacted the farmer profitability and "acceptance" on the demand side, Griliches does not mention weather effects in the text of his work.

5. It should be noted that in the mid-1950s, Griliches did not have access to low-cost computing power to conduct his econometric analysis. His records show calculations made by hand. This helps explain why he sought such parsimonious specification.

6. Griliches relied on a USDA publication entitled "Seed Crops." These data are essentially the same as in USDA (1963).

7. These data were based on USDA (1945, 1949, 1950).

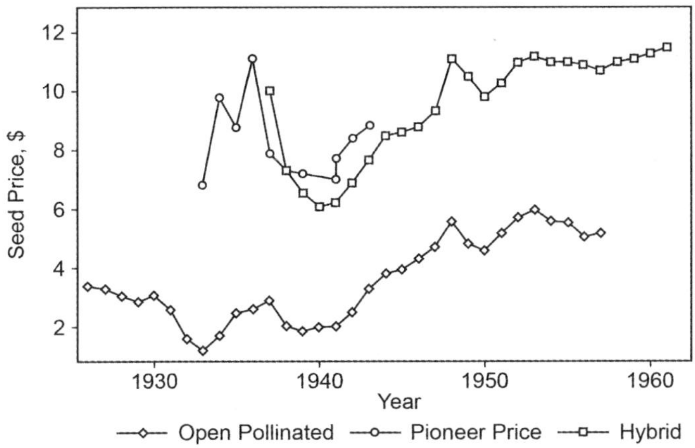

Fig. 3.1 Nominal prices ($) of open-pollinated (OP) and hybrid corn seeds

Sources: Pioneer Company Archives; USDA (1963). Prices paid by farmers for seed: spring season averages, 1926–61: September 15, 1949–61, prices by states and the United States. *Statistical Bulletin* No. 328 (Washington, DC: GPO).

Griliches's pathbreaking work inspired a vigorous scholarly response (see Skinner and Staiger 2007). Sutch (2008, 2011) revisited the early diffusion of hybrid corn, emphasizing the role of adverse weather shocks.[8] Sutch (2008) argued that marketing campaigns and drought stresses (and the 1936 drought in particular) caused farmers to make the costly switch from open-pollinated to hybrid corns. The use of commercial hybrid seed reduced the self-sufficiency of farmers at a time of severe market stress, plausibly increasing risk. Sutch asserted that the early hybrid varieties were not inherently superior to available open-pollinated seeds and that farmers were rightly slow to adopt the expensive seeds in the late 1920s and early 1930s. Sutch (2008, 11) wrote, "During the Depression hybrid seed was selling in Iowa for $6.00 a bushel. Since a bushel of seed would plant two acres, a farmer would have to expect a financial gain of $3.00 an acre to be tempted to pay full price. Expecting no more than 32 cents per bushel for the crop when sold, the advantage of hybrid seed would have had to approach 9 bushels per acre, not the 4–6 seen in the Iowa field tests."[9]

Sutch (2008) argued the adverse weather shocks of the mid-1930s, in

8. Rural sociologists Bruce Ryan and N. C. Gross (1943, 1950) had conducted an earlier study of how Iowa farmers learned about hybrid technologies and how peer effects influenced their adoption decisions. They found that younger and more educated farmers adopted hybrids more readily than older or less educated farmers. They also highlighted the importance of drought conditions on early adoption.

9. Sutch (2008) noted that commodity prices were low and seed was expensive. His analysis did not mention that seed prices were endogenous, set according to market conditions. Nor did he address the subsidies hybrid seed producers gave farmers to adopt hybrids. One strategy seed sellers used to promote adoption was to initially offer farmers enough hybrid seed to plant half a field and take payment as the difference in yields at the end of the growing season.

combination with an intense USDA propaganda campaign, convinced midwestern farmers to adopt the new seed. He noted the conflict of interest that hybrid pioneer Henry A. Wallace faced serving as USDA secretary while retaining ownership of Pioneer Hybrid.[10] Other observers in the 1930s, including the *Chicago Tribune*, were even more critical, arguing the yield-enhancing seed increased crop output at the very time that federal farm programs, run by Wallace, sought to reduce output through acreage restrictions.

Narrative evidence suggests that farmers readily noticed that hybrid corn coped with the dry conditions better than open-pollinated corn planted nearby. As one farmer put it, in these very bad years, the hybrid corn was the last to die (Urban 1975). Singling out the 1936 Dust Bowl drought, Sutch (2011) performed an analysis of hybrid diffusion on state-level data in the Corn Belt in the 1930s and argued that the 1936 drought hastened the adoption of hybrids through learning effects. Sutch was hampered by the lack of comprehensive, geographically decentralized data. He was able to identify records on hybrid and open-pollinated seed productivity only from Iowa. With our new data (or more accurately, newly recovered old data), we seek to address these issues afresh and study hybrid performance from the late 1920s through the 1950s.[11] Combining our data sources allows us to construct a panel of hybrid corn yields, open-pollinated yields, yield differences, hybrid adoption rates, temperature exposure, and precipitation at the crop reporting district (CRD) and year levels for the regions where hybrid seeds first diffused.[12]

3.3 Building Our New Panel Data Set

3.3.1 Data on Hybrid Corn Adoption

The hybrid corn adoption data used in this research project come from unpublished USDA data and notes contained in Zvi Griliches's archival collection held at the Special Collections Library at Harvard University.[13] These data, on the percentage of maize acreage planted in hybrid seed, are available at the level of the CRD. These detailed records—drawn from a grid of roughly nine entries per state—are based on unpublished data from the USDA's Agricultural Marketing Service (AMS). We have recovered these

10. Pioneer was one of the leading commercial seed companies; other leadings hybrid producers at the time included DeKalb, Funk Farms, and Pfister.

11. We thank Richard Sutch for making us aware that the CRD-level diffusion data were available in the Griliches archives.

12. CRDs are relatively equivalent to contemporary agricultural statistics districts.

13. We thank Diane Griliches for allowing access to these materials. We have also sought data at the USDA and AAA collections at the National Archives and the National Agricultural Library.

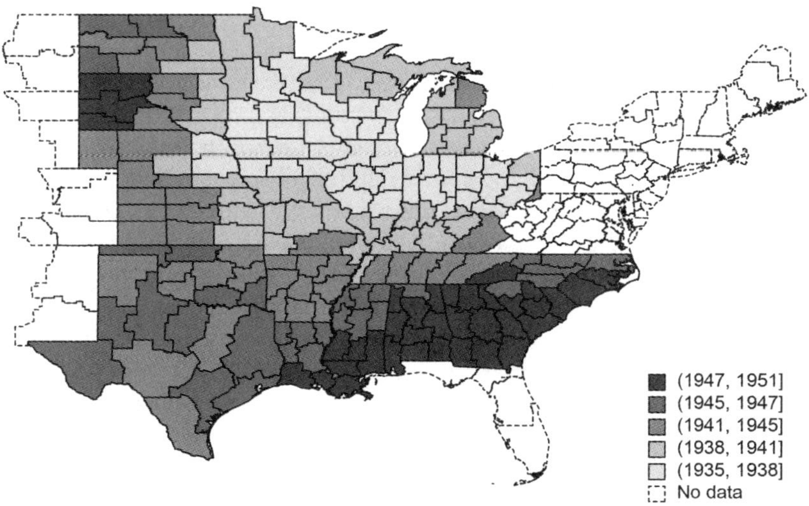

Fig. 3.2 Crop reporting district map, years when hybrid corns exceed 10 percent of planted corn

Source: Compiled from Zvi Griliches's Archival Records.

series for use for the first time since the 1950s. Many previous researchers had to rely on the USDA's state-level data from *Agricultural Statistics*.

The CRD diffusion data were compiled from (1) a set of handwritten spreadsheets for the 1944–55 period; (2) a spreadsheet for Ohio CRDs for the 1935–54 period; (3) very carefully marked diffusion graphs drawn by Griliches's own hand, all in box 58; and (4) typed sheets for all the CRDs in the United States in 1959 in box 60. The graphs indicate the annual rate of diffusion by CRD for each state on a 100-point (or finer) scale covering the period from the first diffusion to 1954/55. The numbers derived from the graphs match exactly those from available nongraphical sources.[14]

The adoption data allow us to define the region of interest for this study. Figure 3.2 visualizes how hybrid corn rapidly diffused across the Corn Belt and the United States in the years following its initial introduction.

3.3.2 Data on Yields of Hybrid and Open-Pollinated Corns

The yield data used in this empirical inquiry come from two primary sources. The first source is the data from experimental farm trials in Iowa

14. We have data for the northeastern states from 1945 on. However, these data do not cover the period of early hybrid adoption in northeastern states. We are seeking to supplement these data but have not been successful in our search for other archival sources. Griliches did collect maps of CRD data from the AAA for the 1938–41 period (box 57). The AAA data have more extensive geographic coverage than the AMS data that Griliches chiefly used. Where there is overlap, the differences are relatively minor.

from 1928 to 1942. These trials, which compared the relative performance of hybrid seeds to open-pollinated seeds, are reported in Zuber and Robinson (1941, 1942). These were the sources that Sutch investigated.[15]

The second source of information on yields is unpublished data held in the Griliches archives. Griliches collected voluminous data on the differential yields achieved by hybrid seed relative to open-pollinated seed. The data, including the results of state yield trials and Agricultural Adjustment Administration (AAA) surveys as well as some yield data, are at the substate level (boxes 57, 60).[16] The CRD data that Griliches actually used in his analysis were derived from AMS studies of "identicals," covering the period from 1939 on (box 59). For early adopting states such as Iowa and Illinois, the series is short because little open-pollinated seed was grown after the mid-1940s. Griliches used the AMS series chiefly in summary form. Note that these data do not allow a direct measurement of the effects of the weather shocks (e.g., droughts) of the mid-1930s. But the Iowa experimental trial yield data do.

Thus Griliches's archival records provide two measures for comparing hybrid and open-pollinated yield differences. For some CRDs, average hybrid and open-pollinated corn yields are available for selected years between 1937 and 1941. Figure 3.3 presents the regions these data cover and the differences between hybrid and open-pollinated yields by quartile.

An alternative measure for the difference between hybrid and open-pollinated yields comes from yield "identicals." These are average differences in hybrid and open-pollinated corn seeds grown on the same farm within a CRD. These "identicals" are more consistently documented in Griliches's archival records. The identical data are reported from 1939 to 1953, have broader geographic coverage than the alternative yield data, and are presented in figure 3.4. With both the seed type yield specific data and "identicals" data, there is a broad geographic coverage. The trade-off with these data is that they cover a time period almost a decade after hybrids had initially entered the market.

The rediscovery and rescue of the yield data, separating open-pollinated and hybrid yields by CRD, again demonstrate the value of archival research. Zvi Griliches was a preeminent researcher who collected and analyzed the pertinent evidence relevant to his study. He knew the importance of making direct comparisons of the yields of corn varieties under comparable settings, at the same time, and in the same place. He collected data from experiment

15. Sutch (2011) described ratios from Iowa corn yield tests as representing all varieties tested. But the data are in fact for *reporting* section varieties, the subset of varieties entered in tests in all three districts in a section. Records from Iowa reports average yield for all and section varieties for 1928–32 ratios for all and the section subset are reported. The average hybrid to openpollinated yield ratio was 1.1069 for all varieties entered but 1.095 for section varieties. A further issue with the test data is that the districts and trial locations change (marginally) over time. How these inconsistencies affect the comparison is unclear, a priori.

16. The substate regions covered do not always translate directly into CRDs.

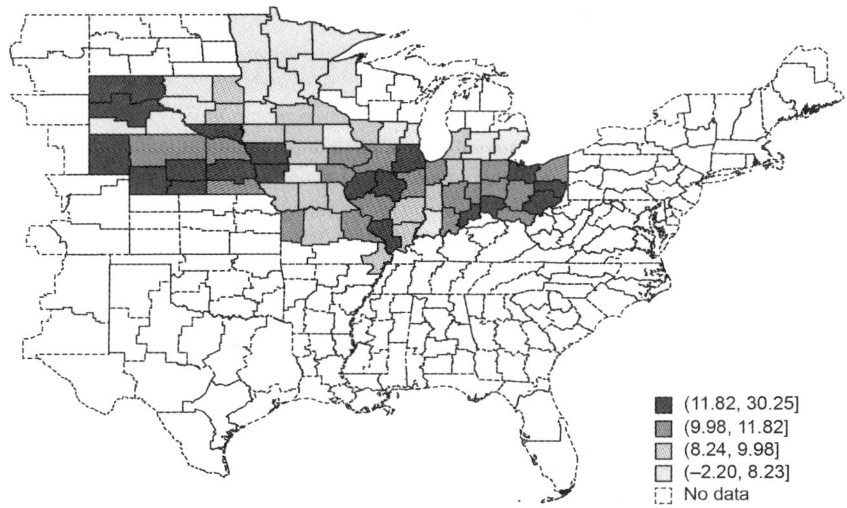

Fig. 3.3 Average hybrid minus average open-pollinated corn yield per acre, quartiles, 1937–41

Source: Griliches's Archival Records.

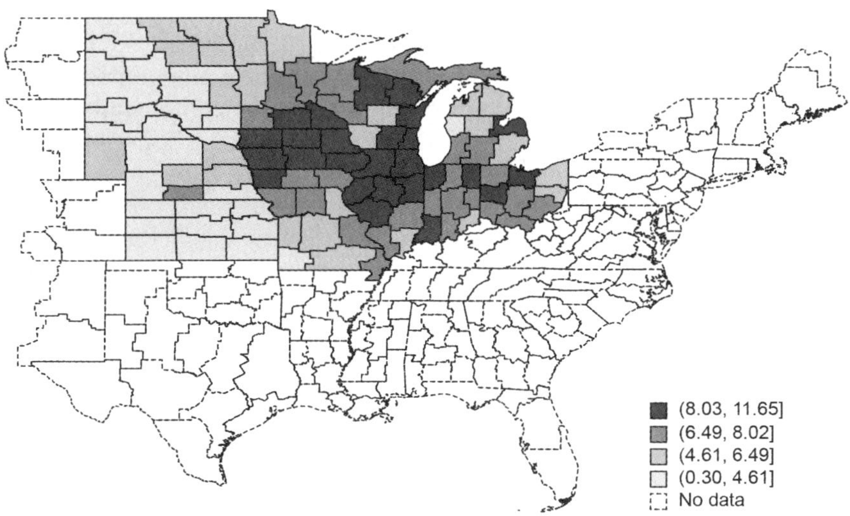

Fig. 3.4 "Yield identicals" per acre, quartiles, 1939–53

Source: Griliches's Archival Records.

station trials and real-world production of identical farms. In his model, the yield difference drove adoption, and hence Griliches sought independent measures of the gap.

3.3.3 Data on Weather

The weather data used in this study also come from two sources. The first is the Palmer drought severity index (PDSI) derived from the Global Historic Climatology Network (Menne et al. 2012). Drought conditions are generally associated with prolonged periods of above-average temperatures and moisture deficits. The PDSI utilized a hydrological accounting model to assess cumulative departure in surface water balance. The PDSI captures drought conditions over a multimonth period and is strongly correlated with observed soil moisture levels (Dai and NCAR 2019).[17]

Extreme deviations in temperature are also often correlated with drought. Therefore, we use temperature and precipitation from Schlenker and Roberts (2009) as our second measure of weather variation.[18] GDDs are aggregations of daily temperature conditions and a common measure utilized in the literature studying climate change and agriculture. Such measures account for deviations in aggregate temperature exposure over a period of time but do not necessarily discern differences between more mild, prolonged temperature spikes and shorter, more extreme temperature spikes. Schlenker and Roberts's information on GDDs is based on the PRISM weather data set.[19] The raw data consist of daily minimum and maximum temperatures as well as total precipitation on a 2.5-by-2.5-mile grid of the continental United States. For each CRD, we use this gridded data to calculate the average daily minimum and maximum temperatures along with total daily precipitation. We then construct GDDs in accordance with agronomically observed heat sensitivity in corn yields, heat in excess of 29°C (Schlenker and Roberts 2009; Schauberger et al. 2017). For each growing season, defined as lasting from April 1 to September 30, we calculate the total number of moderate GDDs and extreme GDDs.

$$(1) \qquad GDD = \begin{cases} \dfrac{T_{max} - T_{min}}{2} - T_{base} & \text{if } \dfrac{T_{max} - T_{min}}{2} > T_{base} \\ 0 & \text{if } \dfrac{T_{max} - T_{min}}{2} \leq T_{base} \end{cases}.$$

The equation above defines a GDD as the average daily temperature calculated between the daily maximum temperature, T_{max}, and daily minimum

17. The index is normalized around 0, with values greater than 0 associated with abnormally wet conditions for a specific region and values less 0 zero associated with abnormally dry conditions for a specific region. Values on the index between −1 and −2 denote mild drought conditions, values between −2 and −3 denote moderate drought conditions, and values less than −3 denote extreme drought conditions.

18. We thank Michael Roberts for recommending that we use this source.

19. See the website at http://prism.oregonstate.edu.

temperature, T_{min}, minus some base temperature, T_{base}. A GDD measures the amount of heat exposure crops receive during a specific day and takes a value of zero for days below T_{base}. Following the example of Schlenker and Roberts (2009), we differentiate between two measures of heat exposure for corn for each CRD for each year from 1920 to 1955 using GDD. We first sum up the number of GDD between 10°C and 29°C during the growing season as moderate GDDs. This calculation assumes a base temperature of 10°C. We sum days with average temperatures in excess of 29°C as extreme GDDs (and assume a base temperature of 29°C in this calculation). In addition to these heat measures, we also total the amount of precipitation during the growing season.

3.4 Empirical Analysis of Our Panel Data Set

3.4.1 Summary Statistics

Tables 3.1 and 3.2 describe the two unbalanced samples constructed for the analysis. In the hybrid and open-pollinated yields sample, most data are for the years 1939 to 1941 and do coincide with the end of the Dust Bowl drought waves. The yield-identical data span from 1939 to 1953 and

Table 3.1 **Summary statistics of hybrid and open-pollinated yields sample**

Variable	Observations	Mean	Std. dev.	Min	Max
Hybrid yield per acre	211	51.430	15.069	13	97.969
Open-pollinated yield per acre	211	40.669	15.854	3.8	90.092
Yield difference	211	10.761	4.8226	2	31.700
Moderate drought, PDSI	211	0.531	0.500	0	1
Extreme drought, PDSI	211	0.289	0.454	0	1
Moderate growing degree days	211	1,773.318	228.086	1,181.853	2,433.123
Extreme growing degree days	211	60.205	29.489	12.248	146.674
Precipitation	211	0.545	0.113	0.254	0.879
Precipitation squared	211	0.310	0.128	0.065	0.773
Year	211	1,939.787	1.103	1,937	1,941

Table 3.2 **Summary statistics of yield identicals sample**

Variable	Observations	Mean	Std. dev.	Min	Max
Yield identical	989	6.029	3.148	0.1	31
Moderate drought, PDSI	989	0.568	0.500	0	1
Extreme drought, PDSI	989	0.267	0.442	0	1
Moderate growing degree days	989	1,581.220	348.529	841.531	2433.123
Extreme growing degree days	989	47.621	38.649	0.721	214.064
Precipitation	989	0.547	0.160	0.197	1.233
Precipitation squared	989	0.325	0.199	0.039	1.519
Year	989	1,944.219	3.876	1,939	1,953

Table 3.3 **Summary statistics of Iowa experimental trials sample**

Variable	Observations	Mean	Std. dev.	Min	Max
Yield ratio, hybrid/open-pollinated	170	114.303	10.585	97.4	153.9
Moderate drought, PDSI	170	0.5	0.515	0	1
Extreme drought, PDSI	170	0.224	0.418	0	1
Moderate growing degree days	170	1,688.860	155.203	1,289.026	2,091.285
Extreme growing degree days	170	57.457	35.310	9.589	206.420
Precipitation	170	0.567	0.088	0.394	0.841
Precipitation squared	170	0.329	0.105	0.155	0.708
Year	170	1,933.565	4.740	1,926	1,941

have broader geographic coverage and more variability in their measures of heat exposure and precipitation. The difference between hybrid corn and open-pollinated yields is on average 10.8 bushels of corn per acre. The yield identical finds a smaller difference of 6 bushels per acre for corns grown on the same farm.

Table 3.3 summarizes the Iowa experimental trial data from 1928 to 1942. The ratio of hybrid corn yields to open-pollinated yields ranges from 97.4 to 153.4 and is on average 114.3. These data suggest that hybrid corn seeds outperformed open-pollinated corns by 14.3 percent between 1926 and 1941. This average is consistent with Griliches's claims.[20]

3.4.2 Empirical Method and Results

To assess the relationship between drought and yield performance of hybrid and open-pollinated corns, we run the following regression specification:

$$y_{it} = \theta_1 MD + \theta_2 ED_{it} + \alpha_i + \gamma_t + \varepsilon_{it}. \tag{2}$$

The variable y_{it} denotes the natural log of the corn yields, yield difference, or yield identical in CRD i in year t. In the Iowa trials data, y_{it}, denote the ratio of hybrid yields divided by open-pollinated yields. These outcomes are regressed on moderate and extreme drought indicator variables constructed from the PDSI, MD_{it}, and ED_{it}. We construct these drought indicators from the average PDSI over the growing season (April–September). Time-invariant effects specific to each CRD are controlled for by using CRD fixed effects, α_i, and a quadratic time trend, γ_t, controls for potential underlying trends, such as concurrent changes in technology, shared across CRDs. Heteroskedastic standard errors, ε_{it}, are clustered at the state (or CRD) level to account for potential correlation in the errors shared across CRDs from the same state.

20. As Sutch (2011) notes, Griliches did not fully credit the yield gaps reports in the Iowa corn yield test data because the farmers engaged in the test program were plausibly not representative of the farm population and achieved yields that were substantially higher than those commonly prevailing.

To assess the effects of heat exposure on the yield performance of hybrid and open-pollinated corns, we run the following linear regression specification:

$$(3) \quad y_{it} = \theta_1 \text{MGDD}_{it} + \theta_2 \text{EGDD}_{it} + \delta_1 \text{Prec}_{it} + \delta_2 \text{Prec}^2 + \alpha_i + \gamma_t + \varepsilon_{it}.$$

The specification using GDDs follows the predominant paradigm used in the agricultural economics literature. The outcomes of interest, corn yields, are regressed on the temperature measures of moderate and extreme GDDs, MGDD_{it} and EGDD_{it}. To address the relationship between corn yields and rainfall we control for growing season precipitation quadratically with PREC_{it} and PREC^2_{it}.

3.4.3 Iowa Experimental Farm Results, 1926–42

Much of the foundational work on developing commercial hybrid corn seeds occurred in Iowa. We use experimental farm data from Zuber and Robinson (1941, 1942) to study the relationship between heat stress and the performance of hybrid corns relative to open-pollinated corns. The data from the Iowa corn yield tests allow us to study hybrid performance when commercial hybrids are introduced and novel. They also let us study hybrid performance during early waves of the Dust Bowl droughts. Figure 3.5 suggests that hybrid yield performance in Iowa was much greater in 1936, a year of extreme Dust Bowl drought, relative to open-pollinated seed lines. It appears the pattern in hybrid to open-pollinated yield ratios starts to shift upward in 1936. Both the floor and average of the ratios also increase until 1942. The last year that yields for open-pollinated corns are reported in Iowa

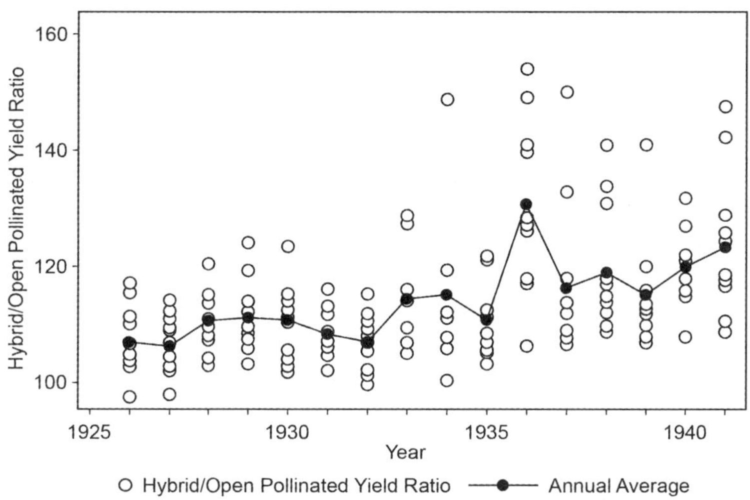

Fig. 3.5 Hybrid to open-pollinated corn yield ratios, Iowa trials data, 1926–41
Source: Authors' tabulation.

Table 3.4 Drought and Iowa hybrid and open-pollinated corn ratios, 1926–41

	Iowa yield ratio (1)	Iowa yield ratio (2)
Moderate drought, PDSI	1.22031	1.64091
	(1.32036)	(1.34115)
Extreme drought, PDSI	6.45368***	14.31629***
	(2.16306)	(3.39188)
CRD fixed effects	Yes	Yes
Quad. time trend	Yes	No
Year fixed effects	No	Yes
Sample	1926–41	1926–41
N	170	170
Adj. R^2	0.251	0.449

Robust standard errors are in parentheses. * $p < 0.10$, ** $p < 0.05$, *** $p < 0.01$.

Table 3.5 Extreme heat and Iowa hybrid and open-pollinated corn ratios, 1926–41

	Iowa yield ratio (1)	Iowa yield ratio (2)
Moderate GDD, 10°–29°C	−0.02343***	−0.02925
	(0.00867)	(0.03604)
Extreme GDD, > 29°C	0.21526***	0.41668***
	(0.04057)	(0.09231)
Precipitation, meters	−12.79066	24.73739
	(65.88781)	(80.88737)
Precipitation^2	14.94762	−17.50428
	(51.77032)	(65.28811)
CRD fixed effects	Yes	Yes
Quad. time trend	Yes	No
Year fixed effects	No	Yes
Sample	1926–41	1926–41
N	170	170
Adj. R^2	0.391	0.488

Robust standard errors are in parentheses. * $p < 0.10$, ** $p < 0.05$, *** $p < 0.01$.

stations is 1942. This is because hybrid corn seed technology had come to dominate corn production in Iowa by that time.

In table 3.4, we regress the Iowa yield ratio against the moderate and extreme drought indicator variables. Moderate drought does not seem to have a differential effect on the relative performance of hybrid seed corn relative to open-pollinated corn. Both specifications (2) and (3) find that extreme drought increases the relative performance of hybrids to open-pollinated corns substantially and indicate that hybrids in the Iowa field trials exhibited some drought tolerance while the open-pollinated corns failed.

In table 3.5, we regress the Iowa yield ratio against the temperature and

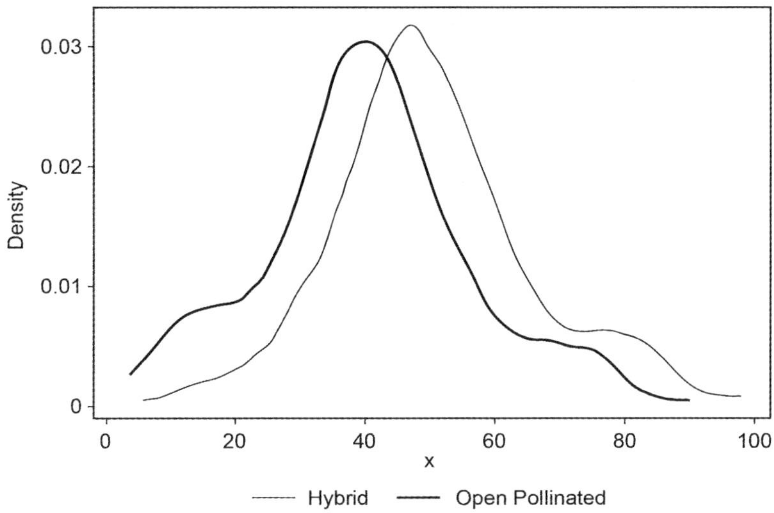

Fig. 3.6 Kernel density plots of hybrid and open-pollinated corn yields, various ranges between 1937 and 1941
Source: Authors' tabulation.

precipitation data. We find results that are consistent with Sutch's (2011) arguments about the role that drought played in diffusion. Specification (2) finds that moderate GDDs decrease the relative performance of hybrids, and the effect is statistically significant at the 1 percent level. A 100-unit increase in moderate GDDs decreases the ratio by 2.3. The statistical significance of the negative effect of moderate GDDs is sensitive to the choice of quadratic time trends or year fixed effects. The coefficients for extreme GDDs show that the relative performance of hybrids increased during periods of extreme heat. In specification (2), a 100-unit increase in extreme GDDs increases the ratio by 21.5 and by 41.7 in specification (3). In both specifications, the coefficients are statistically significant at the 1 percent level. These results support the narrative accounts that hybrid corns performed much better than open-pollinated corns during periods of drought.

3.4.4 Variety-Specific Yield and Yield-Identical Regressions

The yields of hybrid seed corn and open-pollinated seed corn appear to have a fixed gap on average. The kernel density plots of variety-specific yields in figure 3.6 suggest that hybrid seeds shifted the yield distribution to the right. Limiting the sample to CRDs experiencing a drought in a year provides a similar pattern.[21] Figure 3.7 presents kernel density plots of variety-specific yields, and the peaks of the distributions are in similar locations to

21. We define drought as moderate or worse on the PDSI (a value less than −2).

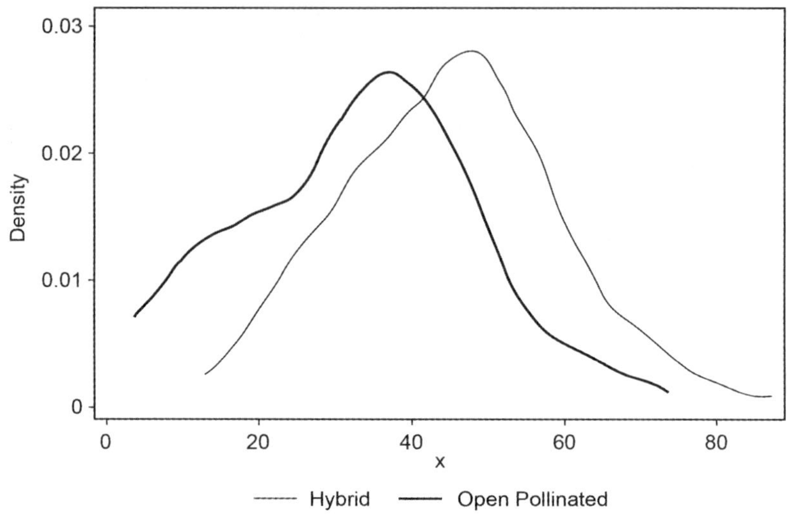

Fig. 3.7 Kernel density plots of hybrid and open-pollinated corn yields under drought conditions, various ranges between 1937 and 1941
Source: Authors' tabulation.

those in figure 3.6. The distribution of yields during droughts shifts probability mass toward the left, but there does not appear to be a stark contrast between the two figures suggesting drought-specific vigor in hybrid seeds relative to open-pollinated seeds. Our regression analysis using yield-specific data further suggests that hybrid seeds were not necessarily drought tolerant relative to open-pollinated seeds.

Columns (1), (2), and (3) in table 3.6 report the effects of moderate and extreme drought on hybrid and open-pollinated corn yields. These yields are averages per acre of specific seeds within each CRD. Column (4) reports the effects of temperature and precipitation on yield "identicals," which is the average difference in hybrid and open-pollinated yields for farms where both seed types were grown. The results from columns (1) and (2) suggest that moderate drought did not strongly reduce hybrid or open-pollinated yields. Extreme drought decreases both hybrid and open-pollinated yields, and the effects are significant at the 5 percent level and below. Nevertheless, the extreme drought indicator variable does not find a strong statistically significant change in either the yield gap or yield "identical" variables. Table 3.7 provides an alternative specification where quadratic time trends are replaced with year fixed effects.

Tables 3.8 and 3.9 report the relationship between GDDs and precipitation on the measures of hybrid versus open-pollinated performance. In table 3.7, the results from columns (1) and (2) suggest that corn yields increase for

Table 3.6 **Regression results, effect of palmer drought severity index drought measures on corn yields, quadratic time trends**

	ln(hybrid yield per acre) (1)	ln(open-pollinated yield per acre) (2)	ln(yield difference) (3)	ln(yield identical) (4)
Moderate drought, PDSI	−0.04312	−0.03525	−0.04267	−0.04166
	(0.03872)	(0.04083)	(0.06336)	(0.04742)
	[0.02462]*	[0.02867]	[0.05517]	[0.04121]
Extreme drought, PDSI	−0.12809	−0.15996	−0.05841	−0.07761
	(0.04796)**	(0.07718)*	(0.09487)	(0.03967)*
	[0.03173]***	[0.04594]***	[0.06713]	[0.05908]
CRD fixed effects	Yes	Yes	Yes	Yes
Quad. time trend	Yes	Yes	Yes	Yes
Sample	1937–41	1937–41	1937–41	1939–53
N	212	212	211	989
Adj. R^2	0.760	0.824	0.471	0.346

Standard errors in parentheses are clustered by state. Standard errors clustered by crop reporting district are in brackets. * $p < 0.10$, ** $p < 0.05$, *** $p < 0.01$.

Table 3.7 **Regression results, effect of PDSI index drought measures on corn yields, year fixed effects**

	ln(hybrid yield per acre) (1)	ln(open-pollinated yield per acre) (2)	ln(yield difference) (3)	ln(yield identical) (4)
Moderate drought, PDSI	−0.02516	−0.03045	0.01987	−0.03732
	(0.03830)	(0.04180)	(0.06250)	(0.04316)
Extreme drought, PDSI	[0.02267]	[0.02857]	[0.06014]	[0.04327]
	−0.09129	−0.14050	0.02762	−0.06327
	(0.07218)	(0.10992)	(0.07857)	(0.05739)
	[0.03849]**	[0.05543]**	[0.07215]	[0.07011]
CRD fixed effects	Yes	Yes	Yes	Yes
Year fixed effects	Yes	Yes	Yes	Yes
Sample	1937–41	1937–41	1937–41	1939–53
N	212	212	211	989
Adj. R^2	0.771	0.824	0.490	0.354

Standard errors in parentheses are clustered by state. Standard errors clustered by CRD are in brackets. * $p < 0.10$, ** $p < 0.05$, *** $p < 0.01$.

both hybrid and open-pollinated corns under moderate GDDs. However, the regression coefficients are essentially the same and suggest that 100 additional moderate GDDs increase corn yields by approximately 6.7 percent (the coefficients are statistically significant at the 1 percent level). Column (3) presents some evidence that hybrid corns perform better relative to open-

Table 3.8 **Regression results, effect of heat stress on corn yields, quadratic time trends**

	ln(hybrid yield per acre) (1)	ln(open-pollinated yield per acre) (2)	ln(yield difference) (3)	ln(yield identical) (4)
Moderate GDD, 10°–29°C	0.00065	0.00066	0.00047	0.00041
	(0.00012)***	(0.00017)***	(0.00035)	(0.00023)*
	[0.00009]***	[0.00012]***	[0.00021]**	[0.00022]*
Extreme GDD, > 29°C	−0.00720	−0.01086	0.00055	−0.00362
	(0.00224)**	(0.00415)**	(0.00382)	(0.00157)**
	[0.00154]***	[0.00243]***	[0.00271]	[0.00136]***
Precipitation, meters	0.05134	1.28623	0.52408	2.16976
	(1.49331)	(1.51984)	(2.34221)	(0.78107)**
	[1.13843]	[1.34332]	[2.03963]	[0.62225]***
Precipitation^2	−0.249615	−1.311283	−0.750254	−1.577880
	(1.22555)	(1.19457)	(2.04302)	(0.49249)***
	[0.93803]	[1.10756]	[1.71669]	[0.42972]***
CRD fixed effects	Yes	Yes	Yes	Yes
Quad. time trend	Yes	Yes	Yes	Yes
Sample	1937–41	1937–41	1937–41	1939–53
N	212	212	211	989
Adj. R^2	0.812	0.876	0.487	0.364

Standard errors in parentheses are clustered by state. Standard errors clustered by CRD are in brackets. * $p < 0.10$, ** $p < 0.05$, *** $p < 0.01$.

Table 3.9 **Alternative specification, regression results, effect of heat stress on corn yields, year fixed effects**

	ln(hybrid yield per acre) (1)	ln(open-pollinated yield per acre) (2)	ln(yield difference) (3)	ln(yield identical) (4)
Moderate GDD, 10°–29°C	0.00068	0.00083	−0.00133	−0.00031
	(0.00061)	(0.00080)	(0.00155)	(0.00022)
	[0.00051]	[0.000579]	[0.00111]	[0.00039]
Extreme GDD, 29°C	−0.00749	−0.01135	0.00308	−0.00250
	(0.00284)**	(0.00477)**	(0.00459)	(0.00151)
	[0.00175]***	[0.00274]***	[0.00338]	[0.00164]
Precipitation, meters	−0.44043	1.018168	−3.05435	1.441982
	(1.62519)	(1.67158)	(2.75151)	(0.86282)
	[1.57646]	[1.84648]	[2.93167]	[0.72885]*
Precipitation^2	0.261705	−0.98615	2.33118	−0.95899
	(1.29111)	(1.24845)	(2.12156)	(0.60857)
	[1.30403]	[1.54234]	[2.45383]	[0.49951]*
CRD fixed effects	Yes	Yes	Yes	Yes
Year fixed effects	Yes	Yes	Yes	Yes
Sample	1937–41	1937–41	1937–41	1939–53
N	212	212	211	989
Adj. R^2	0.810	0.874	0.498	0.366

Standard errors in parentheses are clustered by state. Standard errors clustered by CRD are in brackets. * $p < 0.10$, ** $p < 0.05$, *** $p < 0.01$.

pollinated corns. The yield gap between the two hybrid and open-pollinated varieties increases with additional moderate GDDs, with a 100-unit increase in moderate GDDs increasing the yield gap by 0.48 percent (this result is statistically significant at the 5 percent level when standard errors are clustered at the CRD level). The outcomes for yield "identicals" in column (4) corroborate this result and suggest that a 100-unit increase in moderate GDDs raises the yield gap by 0.42 percent (this result is statistically significant at the 10 percent level under both state and CRD clustered errors).

Extreme GDDs negatively affect the performance of both hybrid and open-pollinated corns. According to columns (1) and (2) in table 3.8, a 100-unit increase in extreme GDDs reduces hybrid corn yields per acre by approximately 51.3 percent and reduces open-pollinated corn yields by 66.2 percent (both coefficients are statistically significant at the 5 percent level). However, there is no statistically significant difference in the gap between the two varieties observed in column (3). According to the yield "identicals" regression in column (4), additional extreme GDDs reduce the performance of hybrids relative to open-pollinated corns. An additional 100 extreme GDDs reduces the yield "identicals" by 30.4 percent (this effect is statistically significant at the 5 percent and 1 percent levels depending on clustering). Only in column (4) does total precipitation during the growing season appear to affect the observed difference in hybrid corn and open-pollinated corn yields. In columns (1) through (3), we find no statistically significant relationship between corn yields and changes in precipitation. This gap appears to be increasing in magnitude until total annual precipitation exceeds 68.7 centimeters, and rainfall decreases hybrid performance relative to open-pollinated corns once total rainfall exceeds 137.3 centimeters. In table 3.9, we present an alternative specification using year fixed effects in place of the quadratic time trends. For all specifications, this change removes all statistical significance associated with moderate GDDs. The statistical significance for the negative effect of extreme GDDs on the yield "identicals" also attenuates. For the hybrid and open-pollinated corn yields, this specification change does not appear to alter yield sensitivity to extreme GDDs. Using year fixed effects does not substantively change the coefficients or statistical significance of extreme GDDs in specifications (2) and (3).

3.5 Conclusion

Our work returning to the original source materials used by Griliches reveals that hybrid seeds increased productivity in corns over a wide range of weather conditions rather than principally during droughts. This finding is consistent with Griliches's assumption that hybrid seed technology increased overall yield potential. We find little evidence that hybrid corn seeds performed differentially better than open-pollinated seeds in periods of drought. If hybrid corns exhibited a unique tolerance toward drought, then we would expect that the difference between hybrid and open-pollinated corn yields to increase in periods of drought. The measures of yield difference suggest that drought conditions decreased the relative advantages of hybrid corns over open-pollinated corns. The evidence using GDDs also does not support the narrative that hybrid seeds outperformed open-pollinated seeds when exposed to extreme temperatures. If anything, the yield advantage of hybrids may have increased during periods of moderate temperatures. Our results indicate that the main benefit hybrid seeds provide in mitigating the adverse effects of drought and extreme temperature is their overall increase in the yield ceiling. This increase in yields cushions the adverse effects of drought.

The arguments made by rural sociologists and historians regarding drought-tolerant hybrids derive from the experiences of early hybrid adopters in Iowa and seem particular to that region during the Dust Bowl. For CRDs in Iowa from 1928 to 1942, extreme temperatures increased the yield performance of hybrid seeds relative to open-pollinated seeds. This evidence is consistent with the claims of Richard Sutch and the rural sociologists regarding the drought-tolerant nature of hybrids. In Iowa, hybrid corns outperformed their open-pollinated contemporaries. The patterns we uncover are consistent with a scenario where farmers' preferences for drought tolerance drove hybrid adoption. Nevertheless, seed producers were introducing a tremendous variety of hybrid seed lines and hybrid varieties marketed outside of Iowa after the period of the Dust Bowl. From the data, it appears these varieties did not exhibit the same drought-resistant characteristics observed in the Iowa experimental field trials.

Appendix

In figures 3.A1 and 3.A2, we plot fitted quadric lines to the data to highlight the relationship between moderate and extreme GDDs and corn yields. We construct estimated hybrid and open-pollinated corn yields using data on harvested corn acreage and output from the National Agricultural Statistics Service's Quick Stats 2.0 program, the yield "identical," and information on share of acreage planted as hybrid corn.[22] This descriptive evidence suggests that hybrid performance increases more under moderate GDDs than open-pollinated corns. It also suggests that the difference in yields is either fixed or decreasing in response to extreme GDDs.

22. The formulas used to construct the data are $\text{Yield}_{op} = \text{Yield}_{total} - \text{Share}_{hybrid} * \text{Identical}$ and $\text{Yield}_{hy} = \text{Yield}_{op} + \text{Identical}$, where Yield_{total} is the overall average yield in a CRD from Quick Stats and Share_{hybrid} is the fraction of acreage planted as hybrid seed.

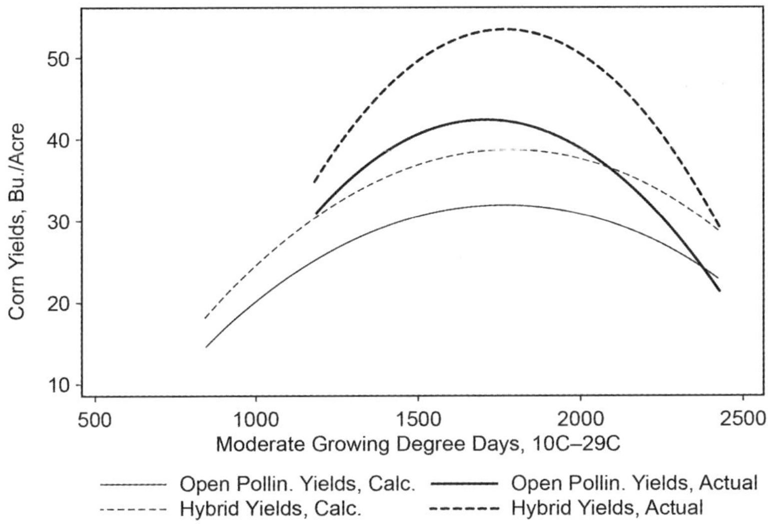

Fig. 3.A1 Moderate GDDs and fitted quadratic lines for constructed and actual hybrid and open-pollinated corn yields

Source: Authors' calculations.

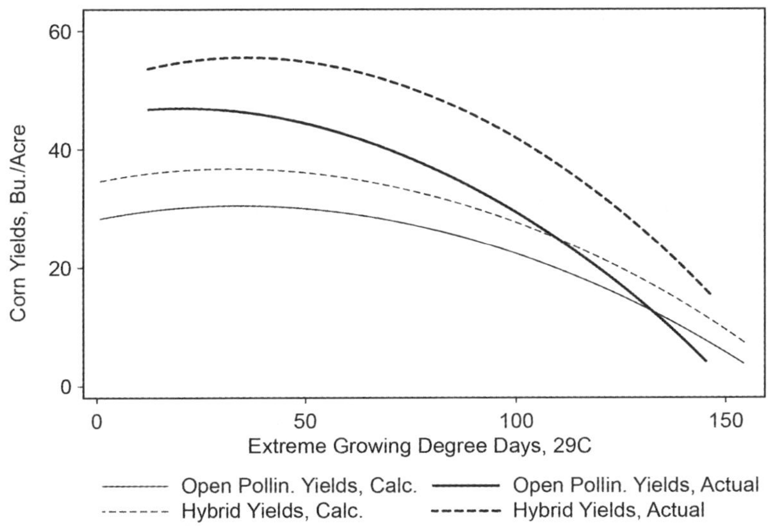

Fig. 3.A2 Extreme GDDs and fitted quadratic lines for constructed and actual hybrid and open-pollinated corn yields

Source: Authors' calculations.

References

Crabb, A. R. 1947. *The Hybrid-Corn Makers: Prophets of Plenty*. New Brunswick, NJ: Rutgers University Press.
Culver, J. C., and J. Hyde. 2001. *American Dreamer: The Life and Times of Henry A. Wallace*. New York: W. W. Norton.
Dai, Aiguo, and National Center for Atmospheric Research (NCAR), eds. 2019. "The Climate Data Guide: Palmer Drought Severity Index (PDSI)." Climate Data Guide, last modified December 12, 2019. https://climatedataguide.ucar.edu/climate-data/palmer-drought-severity-index-pdsi.
Dowell, A. A., and O. B. Jesness. 1939. "Economic Aspects of Hybrid Corn." *Journal of Farm Economics* 21 (2): 479–88.
Fitzgerald, D. 1990. *The Business of Breeding: Hybrid Corn in Illinois, 1890–1940*. Ithaca, NY: Cornell University Press.
Griliches, Z. Papers. Special Collections Library, Harvard University.
———. 1957. "Hybrid Corn: An Exploration in the Economics of Technological Change." *Econometrica* 25 (4): 501–22.
———. 1958. "Research Costs and Social Returns: Hybrid Corn and Related Innovations." *Journal of Political Economy* 66 (5): 419–31.
———. 1960. "Hybrid Corn and the Economics of Innovation." *Science* 132 (3422): 275–80.
———. 1980. "Hybrid Corn Revisited: A Reply." *Econometrica* 48 (6): 1463–65.
Hornbeck, R. 2012. "The Enduring Impact of the American Dust Bowl: Short- and Long-Run Adjustments to Environmental Catastrophe." *American Economic Review* 102 (4): 1477–1507.
Kloppenburg, J. 1988. *First the Seed: The Political Economy of Biotechnology, 1492–2000*. Cambridge: Cambridge University Press.
Menne, M. J., I. Durre, R. S. Vose, B. E. Gleason, and T. G. Houston. 2012. "An Overview of the Global Historical Climatology Network-Daily Database." *Journal of Atmospheric and Oceanic Technology* 29:897–910. doi:10.1175/JTECH-D-11-00103.1.
Olmstead, Alan L., and Paul W. Rhode. 2008. *Creating Abundance: Biological Innovation and American Agricultural Development*. Cambridge: Cambridge University Press.
Pioneer Hi-Bred Company Papers. Special Collections, Iowa State University Library.
Ryan, B., and N. C. Gross. 1943. "The Diffusion of Hybrid Seed Corn in Two Iowa Communities." *Rural Sociology* 8 (1): 15.
———. 1950. "Acceptance and Diffusion of Hybrid Corn Seed in Two Iowa Communities." *Research Bulletin (Iowa Agriculture and Home Economics Experiment Station)* 29 (372). https://lib.dr.iastate.edu/cgi/viewcontent.cgi?article=1386&context=researchbulletin.
Schauberger, Bernhard, Sotirios Archontoulis, Almut Arneth, Juraj Balkovic, Philippe Ciais, Delphine Deryng, Joshua Elliott et al. 2017. "Consistent Negative Response of US Crops to High Temperatures in Observations and Crop Models." *Nature Communications* 8 (1): 1–9.
Schlenker, W., and M. J. Roberts. 2009. "Nonlinear Temperature Effects Indicate Severe Damages to US Crop Yields under Climate Change." *Proceedings of the National Academy of Sciences* 106 (37): 15594–98.
Skinner, J., and D. Staiger. 2007. "Technological Diffusion from Hybrid Corn to Beta Blockers." In *Hard-to-Measure Goods and Services: Essays in Honor of Zvi*

Griliches, edited by Ernst R. Berndt and Charles R. Hulten, 545–70. Chicago: University of Chicago Press.
Sutch, R. 2008. "Henry Agard Wallace, the Iowa Corn Yield Tests, and the Adoption of Hybrid Corn." NBER Working Paper No. 14141. Cambridge, MA: National Bureau of Economic Research.
———. 2011. "The Impact of the 1936 Corn Belt Drought on American Farmers' Adoption of Hybrid Corn." In *The Economics of Climate Change: Adaptations Past and Present*, edited by Gary Libecap and Richard Steckel, 195–223. Chicago: University of Chicago Press.
Urban, N. 1975. "A History of Pioneer's First Ten Years." Unpublished memo, Pioneer Hi-Bred Company Papers. Special Collections, Iowa State University Library.
US Department of Agriculture (USDA). 1945. *Agricultural Statistics*. Washington, DC: Government Printing Office.
———. 1949. *Agricultural Statistics*. Washington, DC: Government Printing Office.
———. 1950. *Agricultural Statistics*. Washington, DC: Government Printing Office.
———. 1963. *Prices Paid by Farmers for Seed: Spring Season Averages, 1926–1961: September 15 Prices, 1949–1961, by States and United States*. Statistical Bulletin No. 328. Washington, DC: Government Printing Office.
Zuber, M. S., and J. L. Robinson. 1941. "The 1940 Iowa Corn Yield Test." *Bulletin P* 1 (19): 1.
———. 1942. "The 1941 Iowa Corn Yield Test." *Bulletin P* 2 (38): 1.

Comment Michael J. Roberts

3.C1 Introduction

Keith Meyers and Paul Rhode consider an iconic and transformational period of time in economic and agricultural history: the adoption and spread of hybrid corn. This topic may seem obscure to some in the discipline, and it would be even more obscure were it not for the famous work of Zvi Griliches (1957), who documented the *S* curve of technological adoption that is now almost universally emblematic of transformation and change. This particular technical change and all that was associated with it—the systematic commercial breeding of seed, massive growth in chemical fertilizer and pesticide applications, and increasing mechanization—mattered tremendously. It marked an acceleration of productivity growth that literally fed the world as its population soared from about 2.3 billion to over 7 billion. Today, we produce over five times as much corn per acre of land as we did before the adoption of hybrid corn (figure 3.C1). Other crops have seen similar advances. With most of the planet's arable land already

Michael J. Roberts is a professor of economics, a fellow of the Economic Research Organization, and an affiliate of the Sea Grant College Program at the University of Hawai'i at Mānoa.
For acknowledgments, sources of research support, and disclosure of the author's material financial relationships, if any, please see https://www.nber.org/books-and-chapters/economics-research-and-innovation-agriculture/comment-yield-performance-corn-under-heat-stress-comparison-hybrid-and-open-pollinated-seeds-during.

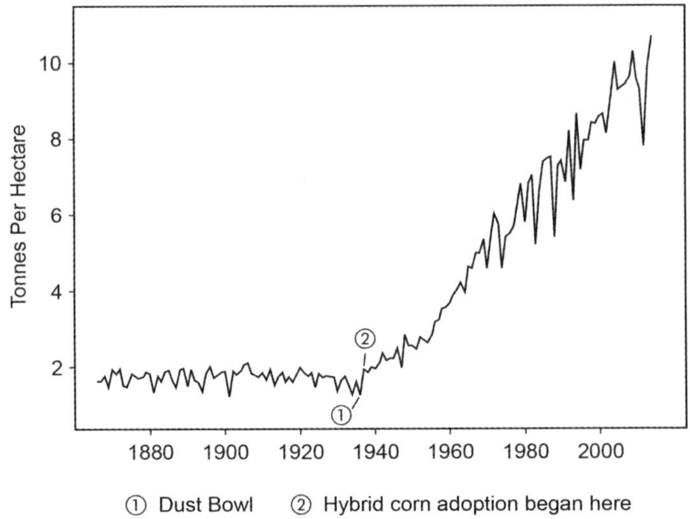

Fig. 3.C1 US corn yields, 1888–2014

Source: These data are from https://ourworldindata.org/crop-yields, which has graphs and data to show a more comprehensive presentation of global agricultural productivity growth. The data can also be obtained from USDA's National Agricultural Statistics Service (https://www.nass.usda.gov).

planted with crops, even 75 years ago, it seems unlikely that, without this innovation, there would be enough arable land on Earth to feed the current population. But with yield growth exceeding average population growth after hybrid corn adoption, the pressure on cropland expansion was greatly subdued. And more importantly, food prices fell even with soaring demand, sparing untold misery. Agriculture today comprises a tiny share of the gross domestic product (GDP) and is a sector we largely take for granted, mainly as a result of this remarkable technical change.

Meyers and Rhode follow up on a hypothesis put forward by Richard Sutch (2011) that hybrid adoption, and the ensuing "Green Revolution" of US and eventually global agriculture, was precipitated by devastating drought and crop failure associated with the Dust Bowl years of 1934 and 1936. In 1937, the year after the most devastating harvest in US history, hybrid corn gained its first substantial foothold, being used for over 40 percent of corn plantings in the most productive counties of Iowa and Illinois. While a productivity boost from hybrids had been discovered much earlier, the gains demonstrated in experimental trials were not enough to justify the high price of commercial seed. The vast majority of farmers still used open-pollinated seed retained from their last harvest, with a much lower opportunity cost.

Evidence presented by Sutch, further buttressed by more formal analyses of Meyers and Rhode, shows that early hybrid varieties not just were higher yielding than open-pollinated corn but also performed relatively well in drought conditions. This boost was especially evident in 1936. When combined with altered expectations stemming from the hottest and driest summers ever experienced in the Corn Belt region, this yield advantage appears to have been enough to entice a substantial number of farmers to purchase and plant commercially bred hybrids for the first time. This likely began in Iowa and central Illinois because these were the most productive areas, and if the productivity boost from hybrids was at least somewhat proportional (as opposed to additive), hybrids would be profitable there first. As adoption spread in later years, it seemed to stretch gradually from higher to lesser productive regions. To a first approximation, more farmers adopted hybrid corn as it became economic to do so, a key theme of Griliches's work.

The main novelty of Meyers and Rhode is their rediscovery and use of a data set collected by Griliches that includes early hybrid adoption rates for each crop reporting district (CRD), each of which is composed of about 10 contiguous counties. Most scholars have used only state-level data, which can obscure important within-state variation in weather, climate, and hybrid adoption rates. These new data allow for considerably more statistical power to test the idea that drought was a catalyst for early hybrid adoption. Meyers and Rhode also employ a formal regression analysis not attempted by Sutch.

A hopeful suggestion of Sutch, and of Meyers and Rhode, is that the prospect of devastating impacts of climate change on US (Schlenker and Roberts 2009) and global (Lobell, Schlenker, and Costa-Roberts 2011) agriculture may induce innovation and productivity growth that far surpasses nearer-term damages, much as the Dust Bowl seems to have done in the late 1930s. Matthew Kahn (2013) has a similar hopeful outlook. While I am generally persuaded by Sutch, Meyers, and Rhode that drought conditions of the late 1930s hastened the early adoption of hybrid corn, it is hard to know by how much this mattered for the long-run productivity trend. And while many point to genetically modified crops—including "drought-tolerant" varieties—as an emerging technology that could aid our adaption to climate change, I think there is good reason to be skeptical that these will impart a second Green Revolution as substantial as the one launched by hybrid corn.

In the next section, I briefly review some technical suggestions, some of which appear to have been adopted by the authors. In the third section, I step back to consider this chapter in the broader context of what we know about potential climate change impacts on agriculture more broadly and why the next Green Revolution, if it comes, may look quite different from the first.

3.C2 Technical Comments

3.C2.1 Measuring Drought

The key prediction variable that Meyers and Rhode considered in their first draft was the Palmer drought severity index (PDSI). This choice of measure is understandable given its prevalence. Drought is a complex thing to measure, for it depends on both supply and demand of soil moisture. Rainfall drives supply, while ground cover (plant type) and vapor pressure deficit (a lack of humidity) drive demand via evapotranspiration and evaporation. The PDSI was the first measure to account for both supply and demand of moisture, and it is based on ground cover of native grasses in Kansas, plus parametric assumptions calibrated by Palmer over 50 years ago (Palmer 1965). Unfortunately, the PDSI just does not predict corn yield outcomes very well.

In our experience, the single weather measure that best predicts corn yields is that of extreme heat—that is, growing degree days (GDDs) above 29°C (Schlenker and Roberts 2009). A quadratic in growing-season total precipitation and/or a measure of precipitation in July and August aid predictions, but only slightly. This extreme heat measure is highly correlated with average vapor pressure deficit in July and August, which suggests that the measure captures drought (Roberts, Schlenker, and Eyer 2013). A physiological model of corn plant growth and seed formation can predict outcomes slightly better, and this model considers the full chronological sequence of weather, corn evapotranspiration, and soil water balance, but it still cannot fully account for the impacts of extreme heat (Roberts et al. 2017). The notable sensitivity to extreme heat is a critical worry about climate change, and it looks like recent crop varieties, almost all genetically modified, are even more sensitive to it (Lobell et al. 2014; Roberts and Schlenker 2014), a point I will return to in the next section.

3.C2.2 Regression Model

Meyers and Rhode developed a model that predicted the log odds ratio of hybrid corn adoption as a function of previous season weather, CRD fixed effects, and a set of baseline 1930 (pre–Dust Bowl) characteristics to account for pre–Dust Bowl trends. The model seems fine as a first cut, but I had some concerns and suggestions. First, while the adoption of hybrid corn is a discrete choice and reversible in principle, the data suggest strongly that the decision is irreversible or nearly so. Thus a change in the adoption share is likely to be permanent. This suggests the use of differences in the log odds ratio instead of levels. Unit root tests could be employed as a formal test between levels and differences, but such tests are notoriously weak, so it may be best to simply report results for both levels and differences and perhaps consider a validation exercise that compares out-of-sample forecasts with actual adoption rates. Another approach may be to use a survival model wherein the dependent variable is the time until *X* percent of acreage in a crop district is planted with hybrid corn. To account for time-varying factors affecting survival time (like exogenous weather), the well-known Cox proportional hazard could be used.

3.C2.3 Zeros and Ones

To calculate the log odds ratio, Meyers and Rhode need to adjust the raw shares of hybrid and open-pollinated varieties, since these shares are often one or zero, and the log is undefined. In their initial paper, they added 0.001 to zeros and subtracted 0.001 from ones. This decision struck me as ad hoc, one that could cause high-leverage outliers that could bias regression results. Instead, it would be preferable to adjust all values by the same constant, not just the zeros and ones, so that the transformation of variables is consistent. This is sometimes called the Haldane-Anscombe correction. Better would be to make the adjustment an estimable parameter. For example, if Y is the share of acreage planted with hybrid corn and a is the adjustment parameter, log odds ratio then becomes $\log(Y + a) - \log(1 - Y + a)$, where a is a parameter to be estimated.

3.C2.4 Spatial Correlation

Climate, weather, soils, and many unobservable factors are all spatially correlated. Thus while weather can be a compelling instrument due to its exogeneity and near randomness in a fixed location when looking over appropriate time scales, regression errors tend to be highly spatially correlated. Meyers and Rhode cluster residuals by CRD, which accounts for serial correlation within CRDs, but residuals will still be correlated, and strongly so, for bordering CRDs. The CRD fixed effects only account for geospatial differences in mean outcomes. Within a year, however, all manner of unobservables and weather anomalies will be similar in nearby areas. Estimated standard errors will be too small. While it can be challenging to get standard errors right, when modeling crop yields in US agriculture, Wolfram Schlenker and I have found that clustering by state gives very similar standard errors as more sophisticated methods, such as Conley's method adapted for panel data.[1]

3.C2.5 Preadoption Productivity Differences

As noted earlier, Meyers and Rhode accounted for preadoption differences in productivity and other factors by interacting year fixed effects with variables from the 1930 Census of Agriculture. This strikes me as a sensible approach. The one concern I have is that a key control variable here is the 1930 yield, which is a rather transitory measure of productivity. It is important to recognize that crop yields vary tremendously from year to year and region to region, largely due to the weather, such that the yield outcome

1. To implement Conley's method adapted for panel data with independent time periods, originally used by Schlenker and Roberts (2009), see the code developed by Thiemo Fetzer (2014) and Solomon Hsiang (2010). The Conley approach may be less appropriate for data with serial correlation, such as hybrid adoption.

from any single year can be a poor reflection of anticipated or expected yield, which presumably drives decision-making. The added variance will likely cause attenuation bias of the control variable coefficient and therefore insufficiently account for preadoption productivity differences. Instead of using the outcome from this one census year, the authors could instead use average annual yield over the decade from 1920 through 1930. Annual-, county-, and crop district–level data are available from the US Department of Agriculture (USDA) to construct such a measure.

3.C2.6 Adoption Model, Yield Model, or Both?

Meyers and Rhode presented results from a regression model predicting hybrid corn adoption as a function of past weather. This approach seems reasonable. These results are nicely complemented by a working paper by Claire Palandri, David Popp, and Wolfram Schlenker (2019), who consider how hybrid corn adoption affected sensitivity to extreme heat, and I shared those preliminary results with Meyers, Rhode, and other attendees at the NBER workshop. Palandri, Popp, and Schlenker present regression results similar to what Meyers and Rhode also exhibit in their chapter. Both works show that hybrid adoption is associated with lower sensitivity to extreme heat, which seems consistent with the idea that the extreme drought conditions of the Dust Bowl years may have been a catalyst for adoption.

A key question I posed at the NBER workshop was whether the apparent drought resistance of early hybrids reflected in the postadoption observational data was the right magnitude to rationally provoke adoption in subsequent years. That is, can the adoption model and yield-response models be reconciled?

This question still needs answering. Careful development of the answer may be complicated. A critical piece concerns the way drought incidence in the 1930s changed expectations for future drought. Broadly speaking, weather looks approximately independent and identically distributed from one year to the next, so a severe drought in one year should not typically lead to altered expectations for the next year. The 1930s, however, were quite different, having several of the hottest and driest years on record in short succession.[2] Perhaps the appropriate way to tie the yield and adoption models together is to consider how much future expectations would have needed to change following the droughts of 1934 and 1936 in order to entice the rational adoption of hybrid corn. Another related question concerns how long diminished expectations would have needed to persist to continue influencing adoption in future years, as drought became less prevalent. Was drought an ongoing impetus for the expansion of hybrid corn or simply a catalyst in the initial year? These questions do not have clear answers, but it would seem that a rolling time series weather forecast, or perhaps even a

2. See the top panel of figure 2 in Palandri, Popp, and Schlenker (2019).

short-window moving average (over, say, a backward-looking five years), might give a reasonable proxy for expected weather.

3.C3 Lessons for Adaptation

While possible links between the Dust Bowl and the emergence of hybrid corn adoption are interesting and should be studied in greater depth, I think we ought to be circumspect about drawing lessons about adaptation to climate change today. The evidence brought to bear so far suggests that the hot and dry Dust Bowl years spurred the initial adoption of hybrid corn in Iowa and Illinois. At the same time, it is hard to see how hybrids would not have emerged anyway, if only a year or two later. Some of the analysis I suggest above may shed greater light on the enduring importance of extreme drought as a motive to adopt new technologies.

The trade-offs associated with the adoption of new crop varieties are quite different today. Many point to new genetically modified crops, and especially "drought-tolerant" varieties, as a viable means of adapting to climate change. The data, however, indicate that today's very-high-yielding corn varieties are *more* sensitive to extreme heat than past varieties (Lobell et al. 2014).

Growing drought sensitivity may be a testament to the success of seed development more generally. At a fundamental level, plant growth is Leontief, limited by the input in least supply. The critical inputs: sunlight, water, and nitrogen, which are the fundamental inputs to photosynthesis. Over time, crops have been bred to take the greatest advantage of the available resources to generate the maximum possible yield in each location (Cassman, Grassini, and van Wart 2010; Wright 2012). In earlier decades, the critical limiting input was nitrogen, which, even in rich soils, naturally occurs in much smaller quantities than became available after the Harbor-Bosch process made chemical fertilizer possible. Crop plants needed to be bred to manage higher nitrogen intake. The plants needed to be able to grow larger, stiffer, and with deeper roots to stand taller in higher planting densities. Thus early crop breeding led plants to have much higher *yield potential*, which crop scientists define as the maximum possible output given available sunlight and water, assuming a sufficient availability of nitrogen and no pest damage. Successive crop varieties were bred to fit the available sunlight and water in each area and to handle massive growth in fertilizer inputs.

Today, nitrogen is almost never the limiting factor; indeed, excess applications, applied just in case moisture and sunlight are sufficiently high, are the key reason for nutrient runoff into streams, lakes, and oceans, causing algae blooms and eutrophication (Babcock 1992; Tilman et al. 2002). And while today's plants have remarkable yield potential, the large plants with deep roots transpire much more water than crops from earlier generations or, for that matter, native grasses that underpin the PDSI. In more recent generations, genetically modified crops have aided management, making it easier to control weeds (glyphosate) and pests (BT strains), thereby helping farmers close the gap between yield outcome and yield potential. As a result, outcomes are more likely to be limited by other essential inputs, especially water. Thus one hypothesis is that earlier generations of hybrids grew yield outcomes mainly by growing yield potential, while later generations grew yield by closing the gap between realized yield and yield potential (Grassini, Yang, and Cassman 2009). Today, for many crops, and especially corn, crop scientists suggest that we are approaching the limit of what is possible (Van Wart et al. 2013).

Despite climate change, the highly productive Corn Belt has, so far, never experienced drought conditions nearly as severe as the 1930s. While the Midwest has experienced warming, it has mainly come during the winter and early spring, while summers have been relatively mild. Crop varieties have been bred to take the greatest advantage of the almost-ideal climate. While mild summers in the Corn Belt region have been a boon to US agricultural production, the region's vulnerability was evident during the unusually hot 2012 season, which came closer to Dust Bowl extremes (Boyer et al. 2013). Projections from climate models suggest that we have been lucky so far to have not experienced more years like 2012. More pointedly, I am not aware of any emerging technology that is likely to change the fundamentals of crop production the way hybrid corn did in the 1930s. At least for some crops, like corn, we are approaching the limits of photosynthesis (Grassini, Yang, and Cassman 2009). This suggests a new Green Revolution will require an altogether different approach.

Climate adversity and associated higher prices might push future innovation. But this will likely take time, as it always has. To my knowledge, at least since the adoption of hybrid corn and the birth of modern agriculture, productivity trends appear roughly linear over time and divorced from obvious inducing incentives, like prices or extreme events (Grassini, Eskridge, and Cassman 2013), with hybrid adoption a very notable exception. If we want to count on the idea that induced innovation will save us from climate change impacts, then I believe we need considerably more evidence to support this hopeful vision. To me, the prospect is daunting because "more than 90 per cent of the calories that feed humanity come from the handful of plants that our ancestors domesticated between 9500 and 3500 BC—wheat, rice, maize (called 'corn' in the US), potatoes, millet and barley" (Harari 2014, 78). Most aspects of our economy—energy, clothing, housing, retail trade, communication—have undergone multiple reinventions in history. Food production has not.

References

Babcock, B. A. 1992. "The Effects of Uncertainty on Optimal Nitrogen Applications." *Review of Agricultural Economics* 14 (2): 271–80.

Boyer, J., P. Byrne, K. Cassman, M. Cooper, D. Delmer, T. Greene, F. Gruis et al.

2013. "The US Drought of 2012 in Perspective: A Call to Action." *Global Food Security* 2 (3): 139–43.
Cassman, K. G., P. Grassini, and J. van Wart. 2010. "Crop Yield Potential, Yield Trends, and Global Food Security in a Changing Climate." In *Handbook of Climate Change and Agroecosystems*, edited by Daniel Hillel and Cynthia Rosenzweig, 37–51. London: Imperial College Press.
Fetzer, Thiemo. 2014. "Conley Spatial Hac Standard Errors for Models with Fixed Effects." trfetzer.com, last updated May 2015. http://www.trfetzer.com/conley-spatial-hac-errors-with-fixed-effects/.
Grassini, P., K. M. Eskridge, and K. G. Cassman. 2013. "Distinguishing between Yield Advances and Yield Plateaus in Historical Crop Production Trends." *Nature Communications* 4 (1): 1–11.
Grassini, P., H. Yang, and K. Cassman. 2009. "Limits to Maize Productivity in Western Corn-Belt: A Simulation Analysis for Fully Irrigate and Rainfed Conditions." *Agricultural and Forest Meteorology* 149:1254–65.
Griliches, Z. 1957. "Hybrid Corn: An Exploration in the Economics of Technological Change." *Econometrica* 25 (4): 501–22.
Harari, Y. N. 2014. *Sapiens: A Brief History of Humankind.* New York: Random House.
Hsiang, Solomon. 2010. "Standard Error Adjustment (OLS) for Spatial Correlation and Serial Correlation in Panel Data in (Stata and Matlab)." Fight Entropy, June 16, 2010. Version 3 update 2018. http://www.fight-entropy.com/2010/06/standard-error-adjustment-ols-for.html.
Kahn, M. E. 2013. *Climatopolis: How Our Cities Will Thrive in the Hotter Future.* New York: Basic Books.
Lobell, D. B., M. J. Roberts, W. Schlenker, N. Braun, B. B. Little, R. M. Rejesus, and G. L. Hammer. 2014. "Greater Sensitivity to Drought Accompanies Maize Yield Increase in the US Midwest." *Science* 344 (6183): 516–19.
Lobell, D. B., W. Schlenker, and J. Costa-Roberts. 2011. "Climate Trends and Global Crop Production since 1980." *Science* 333 (6042): 616–20.
Palandri, C., D. Popp, and W. Schlenker. 2019. "Hybrid Corn Adoption, Average Yields and Heat Sensitivity." Columbia University Center for Environmental Economics and Policy, Working Paper Series.
Palmer, W. C. 1965. *Meteorological Drought.* Vol. 30. Washington, DC: US Department of Commerce, Weather Bureau.
Roberts, M. J., N. O. Braun, T. R. Sinclair, D. B. Lobell, and W. Schlenker, W. 2017. "Comparing and Combining Process-Based Crop Models and Statistical Models with Some Implications for Climate Change." *Environmental Research Letters* 12 (9): 095010.
Roberts, M. J., and W. Schlenker. 2014. "The Evolution of Heat Tolerance of Corn: Implications for Climate Change." In *The Economics of Food Price Volatility*, edited by J.-P. Chavas, D. Hummels, and B. D. Wright, 81–90. Chicago: University of Chicago Press.
Roberts, M. J., W. Schlenker, and J. Eyer. 2013. "Agronomic Weather Measures in Econometric Models of Crop Yield with Implications for Climate Change." *American Journal of Agricultural Economics* 95 (2): 236–43.
Schlenker, W., and M. J. Roberts. 2009. "Nonlinear Temperature Effects Indicate Severe Damages to US Crop Yields under Climate Change." *Proceedings of the National Academy of Sciences* 106 (37): 15594–98.
Sutch, R. 2011. "The Impact of the 1936 Corn Belt Drought on American Farmers' Adoption of Hybrid Corn." In *The Economics of Climate Change: Adaptations Past and Present*, edited by G. D. Libecap and R. H. Steckel, 195–223. Chicago: University of Chicago Press.

Tilman, D., K. G. Cassman, P. A. Matson, R. Naylor, and S. Polasky. 2002. "Agricultural Sustainability and Intensive Production Practices." *Nature* 418 (6898): 671–77.

Van Wart, J., K. C. Kersebaum, S. Peng, M. Milner, and K. G. Cassman. 2013. "Estimating Crop Yield Potential at Regional to National Scales." *Field Crops Research* 143:34–43.

Wright, B. D. 2012. "Grand Missions of Agricultural Innovation." *Research Policy* 41 (10): 1716–28.

4 Local Effects of Land Grant Colleges on Agricultural Innovation and Output

Michael J. Andrews

4.1 Introduction

The US land grant college system is frequently hailed as a major success of agricultural innovation policy (Wright 2012). To be sure, agriculture both in the United States and around the world has become massively more productive over the last 150 years. Moreover, many land grant college towns are now innovation hubs (Harrington and Sauter 2018) and frequently top lists of best places to live (Im 2019). But to what extent are these facts *caused* by the presence of a land grant college, and how much is due to innate location fundamentals?

This question is typically difficult to answer. Simply comparing places with land grant colleges to places without is unlikely to give the true causal effect of a college. Even more frustrating is that it is not clear in which direction this naive comparison is biased. On one hand, land grant colleges were likely established in up-and-coming regions with access to natural amenities like rivers to facilitate transportation and the diffusion of new ideas, suggesting that estimates of the effect of colleges are biased upward. On the other hand, states might choose to locate their land grants close to farmers and far from innovative major cities, implying a downward bias. Indeed, I show

Michael J. Andrews is an assistant professor of economics at the University of Maryland, Baltimore County.

I thank the NBER for financial support. I am very grateful to the participants at the NBER conference "Economics of Research and Innovation in Agriculture," especially to my discussant, Bhaven N. Sampat; to Shawn Kantor and Alex Whalley for sharing data; and to Jeff Furman and Nicolas Ziebarth for thoughtful comments. All errors are my own. For acknowledgments, sources of research support, and disclosure of the author's material financial relationships, if any, please see https://www.nber.org/books-and-chapters/economics-research-and-innovation-agriculture/local-effects-land-grant-colleges-agricultural-innovation-and-output.

that both of these factors were important when states were deciding where to locate their land grant colleges.

To overcome these challenges, I identify cases in which the location of colleges was determined essentially at random. This randomization ensures that estimates of the local effect of land grant colleges represent the true causal effect of the college. More specifically, I use the natural experiments introduced in Andrews (2021b), identifying "runner-up" counties that were strongly considered to become the site of a new college but were ultimately not selected for reasons that are as good as random assignment. As just one example, the location of North Dakota State University was determined by drawing lots, and hence the location of the college site was literally random. The locations of many other land grant colleges, including the University of Maine and the University of Nevada, were decided as the result of particularly close and contentious votes. For still other colleges, such as Iowa State University and the University of Illinois, locations were determined by auction-like processes in which counties submitted bids to receive a new college, and I can compare the bids of the winning and losing finalist sites. While Andrews (2021b) focuses on a broad cross section of different types of colleges, here I narrow the focus to land grant colleges but examine a wide set of agriculture-related outcomes. The first contribution of this chapter is to elaborate on the site selection processes for the land grants, providing detailed narrative evidence about the kinds of decisions made and trade-offs considered when choosing the location for agricultural colleges.

Next, I use the runner-up counties as counterfactuals for locations that received land grant colleges in a differences-in-differences framework and present a number of results. I begin by showing that counties that received a land grant colleges have about 54 log points more patents per year than the runner-up counties after the land grant college is established. I also observe an increase in agriculture-related patents of about 9 log points in land grant colleges relative to runner-up counties after establishing the land grant, although this is imprecisely estimated.[1] I find no evidence that land grant college counties increase the share of county patents belonging to these agriculture-related technology classes. While not precisely estimated, land grant colleges also appear to cause an increase in county population, a factor that is likely to positively affect aggregate invention but may dilute the focus on agriculture.

These results follow a sizable body of research on the local effects of colleges that use patents to proxy for innovation (Andrews 2021b; Hausman 2017; Jaffe 1989; Kantor and Whalley 2014). But patents are less likely to serve as an effective proxy in agriculture than in other sectors because many

1. I use the classification of agricultural patents as defined by Hall, Jaffe, and Trajtenberg (2001). This includes patents that are filed in technology classes related to, for instance, plant and animal husbandry, food, agricultural techniques and processes, and farm machinery like harvesters and combines.

agricultural improvements are not patentable.[2] I make some progress on this issue by using data on the location of origin of new US wheat varieties from a historical US Department of Agriculture (USDA) report (Clark, Martin, and Ball 1922). While the data are much sparser than those for patents, containing information for only 227 new varieties between 1822 and 1922, I find that land grant counties are about five times more likely to introduce a new wheat variety than the runner-up counties after the college is established.

While land grant college counties see sizable increases in local innovations relative to the runner-up counties, they see modest and imprecisely estimated effects on agricultural outcomes, including agricultural yields, total agricultural output, crop output, and livestock production. This overall finding, that land grant college counties have large increases in local agricultural innovation but little increase in local agricultural output, could be interpreted as evidence that either innovations developed at land grant colleges are diffusing to the areas that will use them or innovations developed at land grant colleges are irrelevant for agriculture within the state.

While more study is needed to conclusively distinguish between these interpretations and rule out alternative explanations, the data on wheat varieties (Clark, Martin, and Ball 1922) can again be helpful here. The most commonly planted wheat variety in 1919 was Turkey wheat. Accounting for almost 30 percent of all wheat acreage nationwide, it was likely brought to the United States from Russia in the 1870s by immigrants who settled in rural Kansas. The most commonly planted variety that came from a land grant experiment station was Poole wheat, accounting for about 3.5 percent of national acreage in 1919 after first being documented at the Ohio State University in 1884. On average, in 1919 wheat varieties developed at land grant colleges and their experiment stations tend to be less widely grown than varieties developed elsewhere. This provides some suggestive evidence that land grant innovations may not have been particularly relevant or impactful, although I stress that much more evidence is needed to substantiate this conclusion and to see if it holds for years after 1922.

Are these results on agricultural innovation and performance unique to land grant institutions, or would establishing a college of any type produce similar outcomes? To answer this, I compare my sample of land grant colleges to a sample of non–land grant colleges for which I am also able to identify runner-up locations. These non–land grant colleges do not have the same mandated focus on agricultural research that the land grant institutions do. While measured imprecisely, the estimated increase in local patenting and population is smaller following the establishment of land grant colleges

2. While asexually reproduced plants became eligible for protection under a plant patent in 1930, and both asexually and sexually reproduced plants became utility patent eligible in the late 1980s, none of these methods were available at the time land grant colleges were established. See Moser and Rhode (2012) and Moscona (2019) for studies on the effects of patent protection laws for plants.

than following the establishment of other types of colleges. In terms of agricultural outcomes, the story is less clear: land grant colleges are associated with a larger increase in local agricultural yield relative to other types of colleges but smaller increases in local agricultural output, and in most cases, the magnitudes are small. In short, it is difficult to definitively conclude that land grant colleges play a unique role in promoting local agricultural innovation or output.

Finally, I attempt to get a sense of what drives the observed local effectiveness of land grant colleges. Several pieces of legislation have been passed since the land grant college system was first established in 1862, each of which has affected land grant colleges and their role in agricultural innovation in different ways. One piece of legislation that was particularly important was the Hatch Act of 1887, which established state agricultural experiment stations and provided direct federal funding for agricultural research. The post–World War II era also represented a watershed moment in the federal government's relationship to agricultural research, as exemplified by the Research and Marketing Act of 1946, which reorganized the administration of federal research support and greatly increased the level of federal spending going to land grant colleges. I show that the difference in innovation between college and runner-up counties increases following the passage of these pieces of legislation. This is suggestive evidence that these laws had their intended effect: as funding for agricultural research at land grant colleges increases, these counties indeed produce more innovations. The increase following the passage of these pieces of legislation is larger for land grant colleges than for non–land grant ones, so the effect does not appear to be driven by, for instance, college life cycle effects.

In sum, all these results paint a picture in which explicit funding of agricultural research had large positive effects on the amount of measured agricultural innovation, but there is less clarity regarding how useful these innovations were or how widely they diffused.

This chapter is organized as follows: section 4.2 provides a rich description of the land grant college site selection experiments and describes the sample of colleges used in this chapter, section 4.3 presents the results, and section 4.4 concludes.

4.2 Land Grant College Site Selection Experiments

The main difficulty with attempting to estimate the causal effect of establishing an institution of higher education, including a land grant college, is that these institutions are not located at random. For instance, colleges were often located in up-and-coming areas that were more productive and innovative than other areas in the same state, and so comparing places that get colleges to these other locations will overstate the effect of a college. At the same time, many land grant colleges were located away from productive population centers with the belief that proximity to urban areas would distract students from their learning. On a similar note, state officials frequently wanted to locate public universities close to the geographic center of the state so that they could be equally accessible to all; these concerns often trumped desires to locate colleges in more productive areas. Indeed, many land grant colleges appear to have been located so as to be, as one university president put it, "equally inaccessible from all parts" of the state (Dunaway 1946, 14–15). Hence it is ex ante unclear whether college location decisions are likely to bias estimates of the effects of colleges upward or downward.

To overcome this challenge, I use the data and estimation strategy from Andrews (2021b). More specifically, I examine the historical record to find locations that were finalists to become the site of a new college, similar to the technique used to identify counterfactual locations for large manufacturing plants in Greenstone, Hornbeck, and Moretti (2010). I further restrict attention to cases in which the choice of the winning finalist site is as good as random assignment. I refer to the losing finalists as "runner-up" sites. Andrews (2021b) examines colleges of various types, while in this chapter, my primary goal is understanding the role of land grant colleges.[3]

Andrews (2021b) provides a detailed overview of these natural experiments, including showing that college and runner-up sites are observationally similar prior to establishing the college, showing that college and runner-up sites evolve along parallel trends prior to establishing the college, conducting numerous placebo tests, and describing qualitatively the site selection process, arguing that these decisions were fraught with randomness and unpredictability (see also Andrews 2021a). I therefore take the opportunity here to describe several of these college site selection experiments in more detail than is possible in this other work, providing a deeper understanding of the kinds of historical contingencies at work while referring the reader to Andrews (2021b) for technical details.

I begin with a description of the college site selection process in North Dakota, where the state legislature literally randomly assigned the location of its land grant college, North Dakota State University (NDSU).[4] In an effort to get northern towns to support the move of the Dakota Territory's capital to the south, Territorial Governor Nehemiah Ordway promised other state institutions, including the agricultural college and the state university, to towns in the north. (This push to move the capital would eventually result in the Dakotas splitting into North and South in 1889.) Representatives from the towns of Fargo, Grand Forks, Jamestown, and Bismarck all wanted one of the educational institutions, and despite furious negotiations, they could

3. For the purposes of this chapter, I do not consider historically black colleges and universities (HBCUs) funded under the Second Morrill Act of 1890 as land grants. Reclassifying them as land grant colleges does not qualitatively alter the results.

4. The location of the University of North Dakota was also assigned randomly at the same time and in the same manner; see section 4.2.1.

not be made to agree. Finally, in 1883, with a legislative deadline approaching, the representatives agreed in exasperation to draw lots to allocate the institutions. Fargo won the agricultural college. Seven years later, the school was formally established as the state land grant university (Geiger 1958, 13–27). In the empirical analysis below, I compare Fargo to Jamestown and Bismarck, the runner-up sites, to estimate the effect of the college.[5] One point worth emphasizing is that Jamestown and Bismarck looked very similar to Fargo prior to the establishment of NDSU and, as far as one can ascertain from the historical data, all had the climate, infrastructure, and temperament to successfully support a school. The point is not that the location of NDSU was random but rather that it was random *among the set of finalist locations*. Thus comparing Fargo to only the runner-up sites ensures that the comparison locations are good counterfactuals for Fargo.

Of course, literal random assignment of college sites is rather rare. More common are cases in which states set out a number of criteria that any prospective site must meet and then painstakingly surveyed areas for their suitability. Many "wannabe" locations were eliminated at this stage. Among the remaining candidate locations, a board of trustees or site selection committee would typically meet and debate. Finally, the decision would then come to a vote. These votes were often quite contentious. I consider a candidate location to be as good as randomly assigned if, following this process in which less suitable sites are eliminated, the vote between the winner and the loser is very close. This occurred, for instance, in the cases of the University of Maine (Smith 1979), the University of Nevada (Doten 1924), Clemson University (Reel 2011), and the University of Tennessee (Montgomery, Folmsbee, and Greene 1984).

The University of California, Davis, provides an example of a typical site selection process. Berkeley was originally the location of California's only land grant college, but from the very beginning, critics complained that Berkeley was not climatically representative of the rest of the state and so was a poor site for agricultural research.[6] In 1905, the California state legislature voted to establish a model farm operated independently of the Berkeley campus. The site selection commission was overwhelmed by more than 70 offers from around the state. When narrowing down the sites, the commission set the following criteria: "The farm site should lie within the central portion of the state, in close proximity to a main railroad line, with easy access to good service; its soils should consist largely of medium loam not subject to flooding or under a level; an irrigation system should

5. I do not consider Grand Forks as a runner-up site because it received an institution of higher education of its own. Including the few cases in which the "losing" sites receive a college does not meaningfully alter any results.

6. The original location of California's land grant college was selected because it was close to San Francisco but far enough away to avoid distractions. The trustees settled on Berkeley only after planned land purchases in neighboring counties fell through (Ferrier 1930, 157–214).

already be in place; and the proposed property should be situated within the vicinity of a clean and progressive town. Additionally, [the commission] thought the site ought optimally to represent the state's 'typical' rainfall and general agriculture (i.e., irrigated crops) and avoid extreme heat or other insalubrious conditions" (Scheuring 2001, 18). As this quote demonstrates, representative climatic conditions and infrastructure to support farming were often explicit criteria when deciding land grant locations, providing confidence that winning and runner-up sites are likely similar in terms of their suitability for agriculture. Given the parameters of this refined search, the California commission was left with four finalist locations in Davis, Walnut Creek, Suisun, and Woodland. Although final votes among these finalists are not known, the final meeting to select among these sites dragged on for hours, highlighting just how contentious the decision was. Davis was selected only after speculators tripled the price of land at the commission's first choice. The farm was officially established in 1906 and would become a full-fledged agricultural college in 1921.

The other way in which land grant college sites were often selected was through an auction-like process. Based on the prevailing interpretation of the 1862 Morrill Act, states could use their land grant endowment to fund the operating expenses of agricultural colleges but could not use them for purchasing land or erecting buildings. If a state wanted to create a new agricultural college from scratch, they often solicited bids from localities in the state. I consider the college site to be as good as randomly assigned if candidates' bids are known and the winning bid is very similar to that of the losing bids. These close bidding processes are typically also followed by a contentious vote among a site selection committee. These auction-type processes occurred for schools such as the University of Arkansas (Reynolds and Thomas 1910), the University of Illinois at Urbana-Champaign (Turner 1932; Solberg 1968), Iowa State University (Ross 1958), the Missouri University of Science and Technology (Roberts 1946), and the University of Missouri (Rees and Walsworth 1989; Burnes 2014).

In many cases, the decision of where to locate a college was contentious not only among a site selection committee but also among the residents of the state. The University of Florida provides such an example. In 1905, Florida had eight small institutions of higher education scattered across the state. In an effort to consolidate, the legislature passed the Buckman Act, which closed the existing institutions, reevaluated the best locations, and then reestablished the college at a potentially new site. Gainesville and Lake City quickly emerged as the clear frontrunners to become the new site of the college. Lake City had the added distinction of being the location of the previous Florida Agricultural College. Both Gainesville and Lake City submitted bids of similar amounts, and when it came time for the board of control of the university system to vote on the matter, Gainesville won over Lake City, six to four, following a contentious debate. But as acrimonious as the vote was, it paled in comparison to the views of the citizens of Lake City: as materials from the former agricultural college were being packed to move to their new home in Gainesville, they were done so under an armed guard for fear of rioting (Proctor and Langley 1986, 18–26).

In still other cases, unusual, "fluky" events proved decisive in determining the location of land grant colleges. The establishment of Cornell University (New York's land grant college and the only private land grant institution) provides such an example. What would become Cornell University was originally intended to be located at the People's College in Havana, New York, but the state senator sponsoring the bill suffered an ill-timed stroke, delaying the decision. Later, the legislature was strongly considering placing the college in Ovid when a well-known advocate for the compassionate treatment of the insane died midspeech before the state assembly in Albany. State senators Andrew White and Ezra Cornell were able to use the death to convince the legislature that Ovid should receive an insane asylum instead of a college. Satisfied with the arrangement, Ovid's representatives then decided to support whatever location White and Cornell decided to endorse, creating a dominant legislative coalition (Bishop 1962; Kammen 2003). Even then, the decision was not settled: White and Cornell each wanted to place the college in their hometowns, with White being from Syracuse and Cornell from Ithaca. But Cornell adamantly refused to allow the college to be located in Syracuse because as a young man, he had been "robbed [there] not once but twice" (Kammen 2003, 13); White and Cornell settled on Ithaca instead.

Other colleges provide further examples of serendipity determining a school's location. Louisiana State University moved to Baton Rouge after its prior location burned down, and only a few sites in the state had the infrastructure to take on the school on short notice (Fleming 1936). There are even accounts (possibly apocryphal) that the location of Texas Agricultural and Mechanical University was decided by a poker game (Dethloff 1975, 18)!

Even acts of God intervened to determine college location. In 1885, Arizona's famous (or infamous) "Thieving Thirteenth" legislature met to divvy up the territory's state institutions. The citizens of Tucson had their hearts set on obtaining the state insane asylum when they set off for the legislative assembly in Prescott. But flooding on the Salt River delayed the Tucson delegates, and when they arrived in Prescott, the insane asylum had already been spoken for. The people of Tucson were stuck with the state's land grant college, which became the University of Arizona (Martin 1960, 21–25; Wagoner 1970, 194–222; Cline 1983, 2–4).

As these examples illustrate, the narrative historical record contains rich details about both the locations that received land grant colleges and those that were strongly considered but ultimately did not. Some of these details suggest variation that may be useful for additional analysis. For example, in the case of North Dakota State and the University of Arizona, the "losing towns" that did not receive the land grant college received another type of institution instead. Likewise, in the case of Cornell University, Ovid received an insane asylum in lieu of the land grant college. Syracuse, another runner-up for Cornell University, did not receive any other institution at the time Cornell was established but did receive a university of its own within a few decades. In this chapter, I abstract from these issues, but I discuss them in some detail in Andrews (2021b). Analysis of other types of heterogeneity—such as exploring more finely differences across types of institutions, geography, or other local conditions—may be of interest for future work. All this is possible using the details available in the narrative record.

4.2.1 Non–Land Grant Colleges

Similar strategies can be used to determine runner-up locations for non–land grant colleges as well. As mentioned above, North Dakota drew lots to determine the location of its flagship public university, the University of North Dakota, as well as its land grant college. In the case of the Georgia Institute of Technology, 24 rounds of balloting were required before Atlanta was selected over Macon (McMath et al. 1985, 24–32). For Southern Arkansas University, 8 rounds of balloting were required (Willis 2009, 21–43), and the University of Mississippi took 7 (Sansing 1999, 1–24).[7] Auction-like processes and other "fluky" events are likewise common for the non–land grant colleges.

In sections 4.3.1 and 4.3.2, I use non–land grant colleges as a set of "control institutions" to gain a sense of whether the effects I observe from establishing land grant colleges are caused by policies specifically related to land grants or whether they are common to all institutions of higher education. The appendix lists more details about the sample of non–land grant colleges used in this chapter.

4.2.2 The Sample of Colleges

In total, there are 29 cases in which the site selection decision for a land grant college was as good as random, representing 55 percent of the 53 non-HBCU US land grant institutions. As in Andrews (2021b), all results in this chapter are robust to dropping individual colleges or types of site selection decisions. Table 4.1 lists each of these 29 colleges, the winning county of each, the runner-up counties, and the year in which the college is established.

Table 4.2 presents summary statistics of the land grant college site selection experiments. The median land grant college had 1 runner-up county, with the mean having about 1.5 runner-up counties. The median runner-up site is about 110 kilometers from the college site, although there is consider-

7. Southern Arkansas University actually began as an agricultural school, although it was not a land grant college. The results in this chapter are insensitive to dropping schools like Southern Arkansas or reclassifying them as "land grants."

Table 4.1 **List of land grant college experiments**

	College	County	State	Runner-up counties	Year established
1	Pennsylvania State University	Centre	Pennsylvania	Blair	1855
2	University of California, Berkeley	Alameda	California	Napa; Contra Costa	1857
3	Kansas State University	Riley	Kansas	Shawnee	1863
4	Cornell University	Tompkins	New York	Schuyler; Seneca; Onondaga	1865
5	University of Maine	Penobscot	Maine	Sagadahoc	1866
6	University of Wisconsin	Dane	Wisconsin	Fond du Lac	1866
7	University of Illinois	Champaign	Illinois	McLean; Morgan; Logan	1867
8	West Virginia University	Monongalia	West Virginia	Greenbrier; Kanawha	1867
9	Oregon State University	Benton	Oregon	Marion	1868
10	Purdue University	Tippecanoe	Indiana	Monroe; Marion; Hancock	1869
11	University of Tennessee	Knox	Tennessee	Rutherford	1869
12	Louisiana State University	East Baton Rouge	Louisiana	Bienville; East Feliciana	1870
13	Texas A and M University	Brazos	Texas	Austin; Grimes	1871
14	University of Arkansas	Washington	Arkansas	Independence	1871
15	Auburn University	Lee	Alabama	Tuscaloosa; Lauderdale	1872
16	Virginia Polytechnic Institute	Montgomery	Virginia	Albemarle; Rockbridge	1872
17	North Dakota State University	Cass	North Dakota	Burleigh; Stutsman	1883
18	University of Arizona	Pima	Arizona	Pinal	1885
19	University of Nevada	Washoe	Nevada	Carson City	1885
20	North Carolina State University	Wake	North Carolina	Lenoir; Mecklenburg	1886
21	University of Wyoming	Albany	Wyoming	Uinta; Laramie	1886
22	Utah State University	Cache	Utah	Weber	1888
23	Clemson University	Pickens	South Carolina	Richland	1889
24	New Mexico State University	Dona Ana	New Mexico	San Miguel	1889
25	University of Idaho	Latah	Idaho	Bonneville	1889
26	University of New Hampshire	Strafford	New Hampshire	Belknap	1891
27	Washington State University	Whitman	Washington	Yakima	1891
28	University of Florida	Alachua	Florida	Columbia	1905
29	University of California, Davis	Yolo	California	Solano; Contra Costa	1906

Note: List of land grant college experiments in the sample, along with the winning county and state, the runner-up counties, and the year in which the site selection decision took place.

Table 4.2 **Summary statistics of land grant college experiments**

	N	Mean	SD	Min	Median	Max
# Runner-up counties	29	1.55	0.69	1.00	1.00	3.00
Distance to college	45	150.38	111.88	30.31	109.28	553.35
Year established	29	1877.28	13.28	1855	1872	1906

Note: Number of runner-up counties, average distance from the runner-up counties to the college site, and experiment year for the land grant college experiments in the sample.

able heterogeneity, with the mean runner-up being 150 kilometers away, the farthest runner-up being 550 kilometers away, and the closest runner-up being only 30 kilometers away.

Throughout this chapter, I define the year in which a college is established to be the year in which the college site is selected, as described in the college site selection experiments above. In some cases, this date is not the same as the date at which an institution was formally founded, nor need it coincide with the date at which the college opened its doors. Results are unchanged when using the first year students attended or the first year students graduated as the establishment year. In section 4.3.2, I investigate the importance of other dates in a college's life, such as the year a college began receiving reliable federal research funding. Most of the sample colleges selected their sites and opened their doors in the first decade and a half after the Morrill Land-Grant Act was passed. Two schools were established before the act and obtained land grant status later. Western states typically established their land grant colleges around the same time they obtained statehood, with several states doing so in the 1880s and 1890s. Southern states could not take advantage of the Morrill Act while in rebellion against the US government during the Civil War, so all southern schools in the sample established their colleges in 1869 or later. There is thus substantial temporal variation in the establishment of land grant colleges.

4.3 Results

Figure 4.1 plots four different outcome variables for the land grant and runner-up counties over time. Year 0 is normalized to be the year in which each land grant college is established. In panel (a), I plot logged patenting, in panel (b) logged county population, panel (c) logged agricultural yield (i.e., log(ValueAgr.Output / FarmAcres)), and panel (d) the logged value of all agricultural output. Throughout, all US patenting data come from the data set assembled in Berkes (2018), population data come from the National Historical Geographic Information System (Manson et al. 2018), and all agricultural data come from agricultural censuses, cleaned and compiled by Haines, Fishback, and Rhode (2018). For the popu-

A. log (Patents + 1)

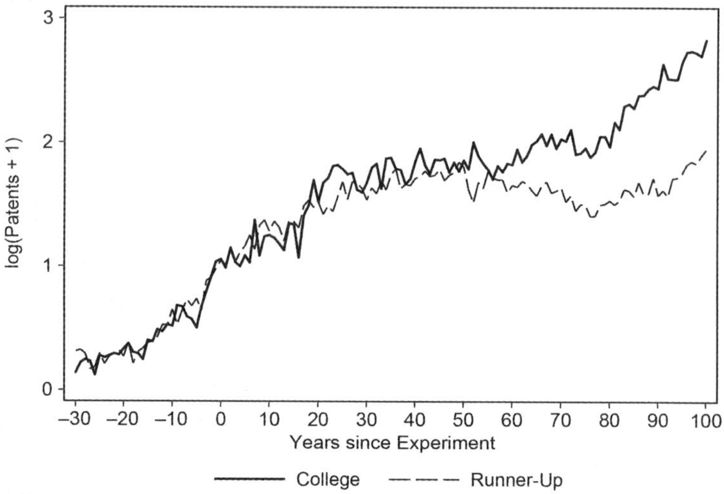

B. log (*Population*)

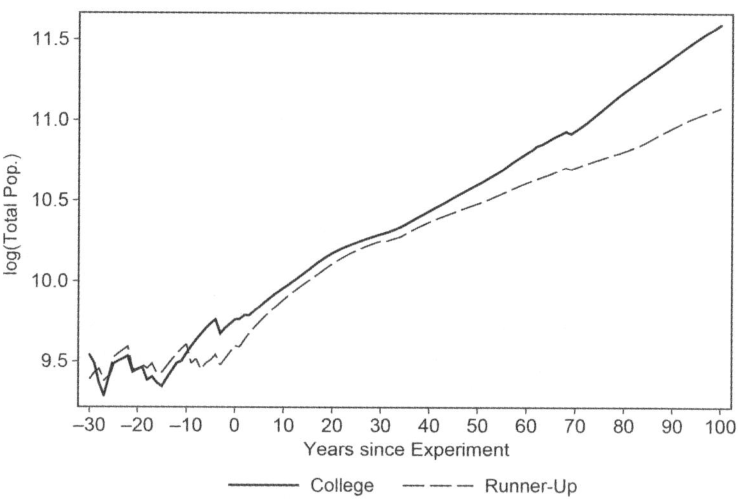

Fig. 4.1 Land grant college counties and runner-up counties

Note: Plots of various outcome variables in land grant colleges (solid lines) and runner-up counties (dashed lines). The *x* axis shows the number of years since the land grant college experiment. The year of the college experiment is normalized to year 0.

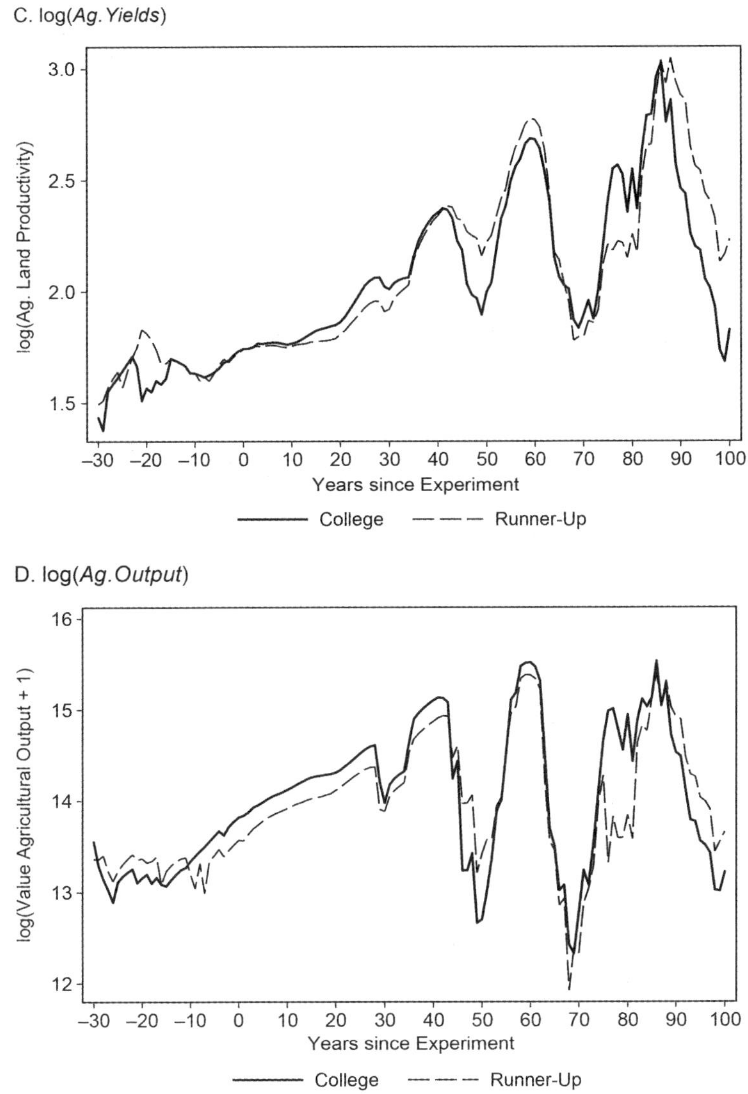

Fig. 4.1 (cont.)

lation and agricultural data that come from federal census data, I linearly interpolate values for all between-census years; unless otherwise noted, results are not sensitive to alternative interpolation approaches or to only using data from census years.

These four pictures tell the main story of this chapter: counties that receive a land grant college see a measurable increase in local invention, especially

Table 4.3 **Differences-in-differences results comparing land grant college counties to runner-up counties**

	log(patents + 1)	log(ag. patents + 1)	Frac. ag. patents	New wheat variety	log(total pop.)	log(frac. urban)
	A. Innovation and population outcomes					
CollegeCounty * PostCollege	0.539**	0.0857	0.00246	0.0168**	0.0966	0.00319
	(0.193)	(0.0624)	(0.0196)	(0.00605)	(0.199)	(0.0304)
PostCollege	0.0970	0.105	0.0228	−0.00711*	0.287	0.0264
	(0.172)	(0.0627)	(0.0147)	(0.00282)	(0.157)	(0.0232)
Num. counties × years	13,141	13,141	9,745	6,639	12,449	9,477
Adj. R^2	0.721	0.314	0.0461	0.00778	0.799	0.702

	log(ag. yields)	log(value agricultural output + 1)	log(value crops + 1)	log(value livestock products + 1)
	B. Agricultural outcomes			
CollegeCounty * PostCollege	0.0998	0.156	0.127	−0.0419
	(0.118)	(0.286)	(0.331)	(0.385)
PostCollege	−0.177*	0.314	0.189	0.628
	(0.0837)	(0.222)	(0.280)	(0.355)
Num. counties × years	11,780	12,190	12,190	12,190
Adj. R^2	0.914	0.923	0.956	0.938

Standard errors in parentheses. * $p < 0.05$, ** $p < 0.01$, *** $p < 0.001$.

Note: Differences-in-differences regression results comparing land grant college counties to runner-up counties before and after establishing each college. Panel (a) uses innovation and population outcomes as the dependent variables. Panel (b) uses agricultural yield and output as the dependent variables. All regressions include county and year fixed effects. Standard errors are clustered at the county level.

after about five decades. There is weak and noisy evidence that land grant colleges also cause increases in population, a major driver of local invention for the larger sample of colleges considered in Andrews (2021b). But the counties that receive land grant colleges see no clear increase in agricultural yield or output relative to the runners-up; while the agricultural measures fluctuate over time, these fluctuations are typically common to both the college and runner-up counties.

Table 4.3 confirms these results in a regression framework. I estimate the simple differences-in-differences model:

$$(1) \quad Y_{it} = \beta_1 \text{LandGrantCounty}_i \times \text{PostLandGrant}_{it} + \beta_2 \text{PostLandGrant}_{it} + \text{County}_i + \text{Year}_t + \varepsilon_{it}.$$

LandGrantCounty_i is an indicator variable equal to one for the counties that receive land grant colleges. $\text{PostLandGrant}_{it}$ is an indicator variable equal to one in years t after the establishment of the college for which county i

was either the winner or runner-up. $County_i$ is a county fixed effect, $Year_i$ is a year effect, and ε_{it} is an idiosyncratic error term. The estimation sample is made up of the college and runner-up counties for all years for which data are available; not all variables are available for all years. In all regressions that follow, I cluster standard errors at the county level.

I estimate effects of establishing a land grant college for a larger battery of outcome variables than I present in figure 4.1. Panel (a) of table 4.3 shows results for innovation and population outcomes. Column (1) confirms the results from panel (a) of figure 4.1: establishing a land grant college causes about 54 log points more patents per year relative to the runner-up counties. Column (2) specifically examines patents classified as agricultural according to the NBER patent classification system (Hall, Jaffe, and Trajtenberg 2001).[8] While the estimated coefficient is positive, it is imprecisely estimated and much smaller in magnitude than overall patenting—at a roughly 9 log point increase in agricultural patents per year. Column (3) shows that there is no significant change in the fraction of agricultural patents in land grant college counties after establishing a new college.[9]

One challenge with measuring agricultural innovation is that many important breakthroughs, particularly the development of new and improved crop varieties, are not patented (Olmstead and Rhode 2008).[10] To provide some insight into the location of nonpatented agricultural invention, I consult a USDA technical report (Clark, Martin, and Ball 1922) that attempts to classify every variety of wheat grown in the United States as of 1920. Crucially, and exceedingly rare among agricultural studies, the authors also provide the histories of each wheat variety, including how, when, and where each variety was developed and/or introduced to the United States. This allows me to investigate the extent to which land grant colleges directly contributed to innovation in the wheat sector. Because individual counties are extremely unlikely to develop more than one variety in a given year, in column (4) I present estimates from a regression in which the outcome variable is an indicator that is equal to one if a county develops a new variety in that year and zero otherwise.[11] Establishing a land grant college has a statistically

8. These correspond to the following US patent classification classes: 8, 19, 71, 127, 442, 504, 43, 47, 56, 99, 111, 119, 131, 426, 449, 452, and 460. The results are robust to using alternative definitions of what constitutes an agricultural patent.

9. This variable is constructed as the number of agricultural patents divided by the number of patents with a known patent class (Marco et al. 2015). Patent class information is still missing for some patents, particularly older ones. This measure is undefined when a county has no patents in a given year and when the class is unknown for all patents in a county in a given year.

10. This is not to say that patent data are irrelevant to an understanding of agricultural innovation, only that patent data alone paint an incomplete picture. Improvements in farm implements and mechanized equipment, often highlighted as vital contributors to American agricultural development (Cochrane 1979; Hayami and Ruttan 1985), were patentable.

11. Note that in contrast to the data on patenting, the wheat varieties data from this report are unavailable after 1922. In ongoing work, I attempt to transcribe more recent USDA reports that contain histories of crop varieties developed in later years and to gather data on yields or

significant increase in the likelihood of introducing a new variety, on the order of 2 percent. Given that the baseline probability of introducing a new wheat variety in a given year for this sample of counties is about 0.4 percent, counties that receive a land grant college are about five times more likely to introduce a new variety after the college is established.

Consistent with panel (b) of figure 4.1, column (5) shows that establishing a land grant college is associated with a positive but statistically insignificant increase in total population of about 10 log points. The fraction of the county population living in urban areas, shown in column (6), is also positive but statistically insignificant and is close to zero in magnitude.

In panel (b) of table 4.3, I show results for various agricultural outcomes. In column (1), I show that establishing a new college has no statistically significant effect on agricultural yields, although the coefficient is positive and nontrivial in magnitude, equal to a roughly 10 log point increase in agricultural yield relative to the runner-up counties. One issue with yield as an outcome variable is that it is defined as the value of agricultural output divided by agricultural land, and establishing a new college may affect both the numerator and the denominator. In particular, a successful land grant college may induce more marginal land to come into agricultural production, decreasing yield while increasing output. In columns (2) through (4), I estimate the effect of establishing a land grant college on several output measures: the total value of agricultural output, the value of crop output, and the value of livestock produced. In all cases, establishing a land grant college has statistically insignificant effects, although the effect is positive and sizable in magnitude for agricultural output and crop output.[12] This suggests that the land grant counties are increasing the amount of agricultural land relative to the runner-up counties, consistent with untabulated results on the amount of improved farm acreage.

In a related paper, Kantor and Whalley (2019) conclude that land grant colleges cause an increase in the value of agricultural output in the areas closest to the college. It is worth exploring why the conclusions in panel (b) of table 4.3 differ from those in Kantor and Whalley (2019). First of all, the two studies use different samples of colleges. My sample consists of all land grant colleges for which I can identify a runner-up location, while Kantor and Whalley (2019) focus on land grant colleges in the Northeast, Midwest, and Texas. However, even when restricting attention to the land grant colleges in the states studied by Kantor and Whalley (2019), I find results

other measures of quality for the different varieties. I thank Paul Heisey for pointing out the existence of these later reports and discussing their potential usefulness for research on the geography of invention.

12. The agricultural results here present one case in which interpolation meaningfully alters point estimates. When using only data from agricultural census years, the coefficients for agricultural yield, agricultural output, and crop output are all smaller in magnitude, and the coefficient on agricultural output becomes negative. These results are available upon request.

similar to those in table 4.3, and if anything, the coefficient on agricultural output is even closer to zero in magnitude; these results are available upon request. The most important difference is that the two studies ask subtly different questions. The independent variable in Kantor and Whalley (2019) is the distance from each county (not just the runners-up) to the land grant college interacted with a year fixed effect, whereas I compare the land grant college counties only to the runner-up counties. While Kantor and Whalley (2019) ask how agricultural output decreases with distance from a land grant college, I compare locations that would have been equally suitable sites to conduct agricultural research and see how agricultural outcomes change when one of these locations gets a land grant college. If land grant colleges are indeed located in the areas most suitable for agriculture, as the discussion in section 4.2 suggests, with surrounding areas less suitable for agriculture and likely less able to take advantage of agricultural innovations, then we should expect to see a negative gradient of agricultural output with distance, as documented in Kantor and Whalley (2019).[13] It should also be noted that I find similar dynamics to Kantor and Whalley (2019): as shown in panel (d) of figure 4.1, the difference between the land grant college counties and the runner-up counties is largest in the earliest decades after a college is established before shrinking to virtually nothing. In contrast to Kantor and Whalley (2019), however, this difference is small in magnitude and statistically insignificant.[14]

How to interpret the large positive coefficients for local innovation outcomes and small-in-magnitude and statistically insignificant coefficients for agricultural outcomes? One interpretation is that the agricultural innovations documented in panel (a) of table 4.3 successfully diffuse throughout the land grant college's state, so the county from which these innovations originated saw little benefit from them relative to the otherwise similar runner-up

13. In the online appendix to Kantor and Whalley (2019), the authors conduct a robustness test using runner-up counties (see their appendix section 4 and tables A2 and A9). As explained in Andrews (2020b), the Kantor and Whalley (2019) runner-up counties include those from low-quality site selection experiments. Additionally, in some cases, I am able to identify additional runner-up sites not used in Kantor and Whalley. The sample of runner-up locations in Kantor and Whalley therefore differ slightly from the sample used in this chapter. In their specification, Kantor and Whalley use the distance from the runner-up counties to each county interacted with year fixed effects as additional independent variables. They show that while the value of agricultural output decreases with distance to the land grant experiment station, it increases with distance to the runner-up counties. Note that this is different from the analysis I conduct in this chapter.

14. Kantor and Whalley (2019) also find significant declines in the value of agricultural output with distance from the land grant for six decades, whereas in panel (d), the difference between land grant and runner-up counties closes after about five decades. Other differences between the studies may explain this discrepancy. As noted above, the two studies use a different sample of colleges. Kantor and Whalley use the passage of the Hatch Act in the 1880s as their date of treatment, whereas I use the establishment of the college (I examine the effects of the Hatch Act in section 4.3.2). And Kantor and Whalley include a number of 1880 county characteristics interacted with year effects as additional control variables.

Table 4.4 **Share of wheat varieties from land grant research**

	Share varieties		Share acres	
	Land grant research	Land grant counties	Land grant research	Land grant counties
Post Morrill Act	0.303	0.355	0.097	0.131
Post Hatch Act	0.347	0.389	0.113	0.166

Note: Columns (1) and (2) list the share of new wheat varieties introduced since the passage of the Morrill Act in 1862 (row 1) and the passage of the Hatch Act in 1887 (row 2). Column (1) shows the share of varieties introduced as a result of land grant college research. Column (2) includes any varieties introduced in land grant college counties, regardless of whether they were the result of programmatic research. Columns (3) and (4) do the same but weight each variety by acreage planted.

counties. Alternatively, the results could be interpreted as evidence that the innovations developed in land grant college counties are irrelevant to agricultural production in the state, or that the agricultural outcome measures are mismeasuring true agricultural productivity.

As a first pass to addressing this question, I again turn to the data on the introduction of new wheat varieties from Clark, Martin, and Ball (1922). In addition to detailed histories of each variety, the report contains results from a 1919 survey of the total national acreage planted in each wheat variety. By comparing acreage planted of varieties developed at land grant sites to those developed elsewhere, I get a sense of whether land grant varieties tended to diffuse widely by 1919. I restrict attention only to varieties introduced since the passage of the Morrill Land-Grant Act in 1862 to avoid counting varieties from before the land grant system could have had any effect.

I present results in table 4.4. In column (1), I count all varieties that Clark, Martin, and Ball (1922) indicate were introduced as a result of research at land grant colleges or state agricultural experiment stations.[15] About 30 percent of all new varieties introduced between 1862 and 1919 came from land grant research. In column (2), in addition to the varieties attributed to land grant research in column (1), I include any varieties introduced in a county that had a land grant college, even if the land grant site was not explicitly mentioned in the varietal history. Including these additional varieties increases the share of varieties from land grant college counties to about 36 percent of all new varieties. In columns (3) and (4), I calculate the national acreage planted of varieties developed at land grant sites. Varieties

15. In the calculations, I include wheat varieties developed outside the United States as long as Clark, Martin, and Ball (1922) can identify the location within the United States at which the variety is first introduced. Many (although not all) of the varieties attributed to land grant research were initially developed outside the United States, lending support to the claims in Alston (2002) and Maredia, Ward, and Byerlee (1996) that federal support of agricultural innovation generated sizable international spillovers.

introduced as a result of land grant research account for only 10 percent of planted acreage, and all varieties from land grant counties account for 13 percent of acreage. Comparing the number of varieties introduced to the acreage results suggests that land grant research produced varieties that were, on average, less useful for American farmers. In row (2), I repeat the exercise but keep only varieties introduced since the passage of the Hatch Act in 1887, which established and provided federal funding for agricultural experiment stations. When restricting attention to this period in which land grant research was on an even firmer financial footing and was conducted in a larger number of geographic locations, land grant colleges account for a slightly larger share of both varieties and acreage (35 percent and 11 percent, respectively). This is also true when including all varieties introduced in land grant counties (39 percent of varieties and 17 percent of acreage).

From these results, it appears that land grant colleges played an outsized role in discovering and inventing new wheat varieties, although varieties developed at land grant locations were less widely planted on average than varieties grown elsewhere. This suggests that land grant colleges may not cause much of an increase in local agricultural yield and output because the agricultural innovations they produce are of low quality or useful for only a small constituency.

I stress that this conclusion is highly preliminary and suggestive, and several caveats are in order. First, Clark, Martin, and Ball (1922) may have been more likely to uncover information on low-quality varieties when they were developed at land grant sites, and so their data may suffer from survivorship bias. Additionally, it is possible land grant colleges played a larger role in the development of different species of crops or in the development of farm machinery, or that their role qualitatively changed in recent decades; additional USDA reports would be particularly useful to address these issues. It is also likely that land grant colleges played a substantial role in promoting the diffusion of wheat varieties developed elsewhere. Indeed, several of the descriptions of varieties indicate that agricultural experiment station researchers scoured the country to discover varieties developed by obscure farmers.[16] Much more work is needed to conclusively determine why land grants appear to have a large positive local effect on innovation but little effect on local agricultural output and yield.

4.3.1 Comparing Land Grant Colleges to Other Types of Colleges

Is there something "special" about the land grant college program, or would the observed positive effects on innovation be observed anytime an

16. As one example, the Wyandotte variety was discovered by researchers from the Ohio agricultural experiment station at Columbus being grown on a farm in Nevada, Ohio, although the variety's exact origins remain a mystery. The Indiana agricultural experiment station in Bloomington frequented Everitt's O.K. Seed Store in Indianapolis to learn about new varieties from across the country.

institution of higher education is established? To answer this question, I use data from all college site selection experiments, not just the land grants.

Figure 4.2 plots the difference between college and runner-up counties separately for land grant and non–land grant colleges for the same four outcome variables as in figure 4.1. Both types of colleges had small and largely constant differences prior to the colleges being established.[17] Both types of colleges exhibit an increase in patenting and population after establishment, although at different rates. In particular, while the non–land grant college counties see almost immediate increases in local population relative to their runner-up counties, the land grant college counties see large increases in population only after about seven decades. The pictures for agricultural yield and output are less clear, with particularly large fluctuations for land grant colleges but no obvious trend.

I next test the difference between the types of colleges more formally in a triple differences framework. I estimate

$$(2)\quad Y_{it} = \beta_1 \text{CollegeCounty}_i \times \text{PostCollege}_{it} \times \text{LandGrant}_i + \beta_2 \text{CollegeCounty}_i \times \text{PostCollege}_{it} + \beta_3 \text{LandGrant}_i \times \text{PostCollege}_{it} + \beta_4 \text{PostCollege}_{it} + \text{County}_i + \text{Year}_t + \varepsilon_{it},$$

where CollegeCounty_i is a dummy equal to one if county i ever receives a college of any type, PostCollege_{it} is a dummy equal to one in years t after the establishment of the college for which county i was either the winner or runner-up, and LandGrant_i is a dummy equal to one if i was either the winner or runner-up for a land grant college.

I present results in table 4.5 for the same outcome variables as measured in table 4.3.[18] The variable of interest, $\text{CollegeCounty}_i \times \text{PostCollege}_{it} \times \text{LandGrant}_i$, should not be interpreted as causal, since colleges are not randomly assigned to be either land grants or other types of institutions. And the triple interaction term is rarely statistically significant, which is not surprising given the relatively small number of college experiments. Nevertheless, the coefficients suggest an interesting pattern. After establishing a land grant college, the growth in patenting in the college counties is about nine log points smaller than the growth in patenting after establishing a non–land grant college. Agricultural patenting also increases by less after establishing a land grant college, although the coefficient is close to zero in magnitude.

17. In all cases, I fail to reject the null hypothesis of parallel pretrends for both the land grant and non–land grant colleges; results are available upon request. The plotted figures can be misleading in the earliest years, since data are not available for all colleges three decades before the college establishment date.

18. Results comparing land grant to non–land grant colleges are similar when restricting the sample of non–land grant colleges to include only public colleges (typically flagship state universities that are not also land grant colleges, such as the University of North Dakota), although the smaller sample of colleges results in less precise estimates; these results are available upon request.

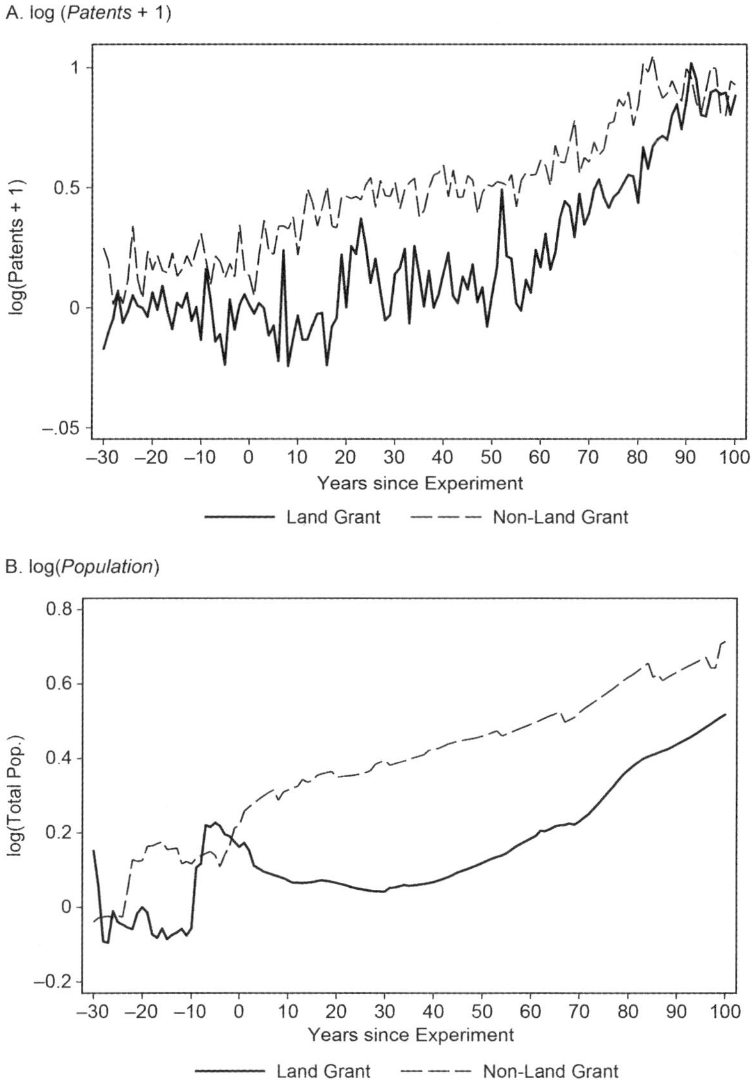

Fig. 4.2 Difference between college and runner-up counties for land grant colleges and non–land grant colleges

Note: Plots of the difference between college and runner-up counties for various outcome variables for land grant colleges (solid lines) and non–land grant colleges (dashed lines). The x axis shows the number of years since the land grant college experiment. The year of the college experiment is normalized to year 0.

C. log (*Ag. Yields*)

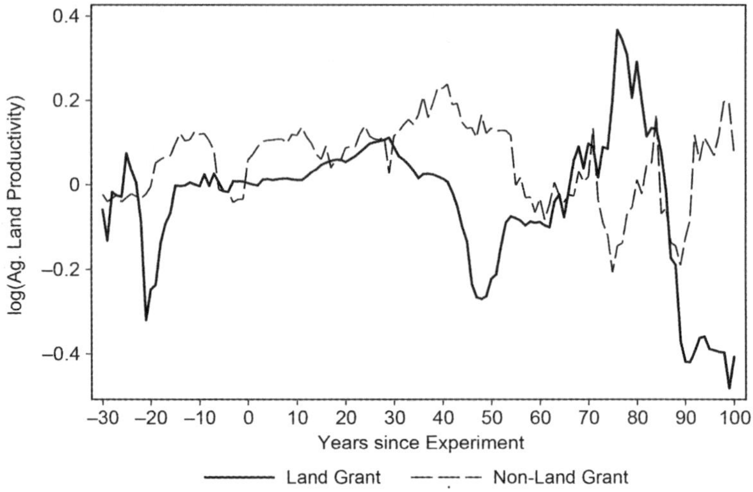

D. log(*Ag. Output*)

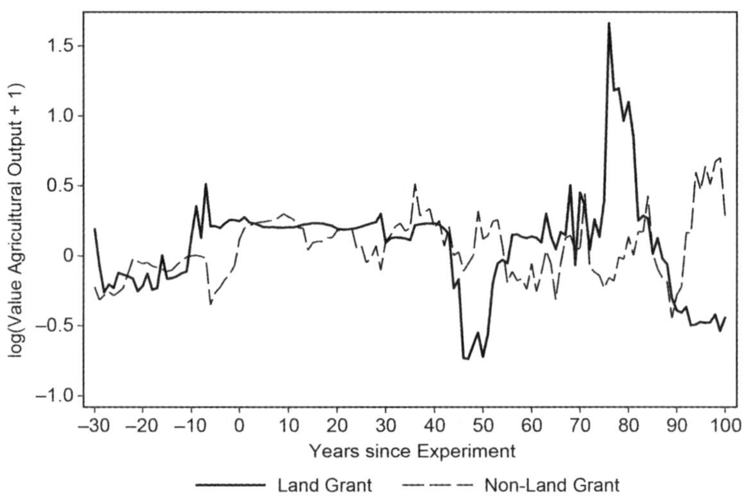

Fig. 4.2 (cont.)

But when focusing on non-patent-based agricultural innovations, land grant colleges do have a larger effect than the non–land grant colleges: the increase in the likelihood of introducing new wheat varieties is 2 percent more in college counties after establishing a land grant college than after establishing a non–land grant institution, an effect that is statistically significant at the 10 percent level. Land grant colleges are also associated with less population growth and urbanization than the non–land grant colleges. Agricultural

Table 4.5 **Triple differences results comparing the LandGrant to non–LandGrant colleges**

	log(patents + 1)	log(ag. patents + 1)	Frac. ag. patents	New wheat variety	log(total pop.)	log(frac. urban)
	A. Innovation and population outcomes					
College * PostCollege * LandGrant	−0.0934	−0.00639	0.00930	0.0157*	−0.385	−0.0616
	(0.263)	(0.0757)	(0.0257)	(0.00640)	(0.262)	(0.0433)
CollegeCounty * PostCollege	0.634***	0.0926*	−0.00842	0.00118	0.487**	0.0649*
	(0.183)	(0.0426)	(0.0170)	(0.00206)	(0.164)	(0.0310)
PostCollege * LandGrant	0.209	0.0798	0.0129	−0.000826	0.216	0.0438
	(0.182)	(0.0570)	(0.0172)	(0.00164)	(0.182)	(0.0267)
PostCollege	−0.126	−0.00841	0.00906	−0.00103	0.00980	−0.00970
	(0.107)	(0.0333)	(0.0116)	(0.00168)	(0.0966)	(0.0164)
Num. counties × years	34,911	34,911	24,115	17,760	33,541	25,601
Adj. R^2	0.724	0.297	0.0527	0.00408	0.803	0.734

	log(ag. yields)	log(value agricultural output + 1)	log(value crops + 1)	log(value livestock products + 1)
	B. Agricultural outcomes			
College * PostCollege * LandGrant	0.0538	−0.0462	−0.0544	−0.146
	(0.144)	(0.366)	(0.432)	(0.472)
CollegeCounty * PostCollege	0.0337	0.203	0.182	0.123
	(0.0985)	(0.219)	(0.275)	(0.265)
PostCollege * LandGrant	−0.177*	−0.0331	−0.0922	0.497
	(0.0751)	(0.199)	(0.245)	(0.265)
PostCollege	0.108	0.227*	0.157	0.0103
	(0.0555)	(0.0953)	(0.121)	(0.119)
Num. counties × years	32,092	33,312	33,312	33,312
Adj. R^2	0.918	0.926	0.966	0.947

Standard errors in parentheses. * $p < 0.05$, ** $p < 0.01$, *** $p < 0.001$.

Note: Triple differences regression results comparing college counties to runner-up counties before and after establishing each college for LandGrant and non–LandGrant colleges. Panel (a) uses innovation and population outcomes as the dependent variables. Panel (b) uses agricultural yield and output as the dependent variables. All regressions include county and year fixed effects. Standard errors are clustered at the county level.

yield appears to increase more in counties that receive a land grant college than in counties that receive other types of colleges, but if anything, land grant colleges see worse outcomes in terms of total agricultural output, crop output, and livestock.

The coefficient on $\text{CollegeCounty}_i \times \text{PostCollege}_{it}$ measures the effect of establishing non–land grant colleges and shows that these other types of institutions also generate sizable increases in local patenting and agricultural patenting and create positive but statistically insignificant and small-in-magnitude increases in agricultural output. Unlike the land grant colleges, the non–land grant colleges create large increases in local population and statistically significant increases in urbanization. The coefficient on $LandGrant_i \times PostCollege_{it}$ measures how the land grant runner-up counties perform after establishing a land grant college relative to the non–land grant runners-up after establishing a non–land grant college and is thus a plausible measure of spillovers from land grants. The coefficient is negative for agricultural yield, agricultural output, and crop output, although it is positive for all measures of innovation. This calls into question whether the land grant colleges were more effective at generating innovations that diffused throughout their states than were other types of colleges. Conclusions about spillovers and diffusion should be made with caution, however, since the non–land grant runner-up counties may be exposed to innovations from a nearby land grant college, and vice versa. A full exploration of these issues is beyond the scope of this chapter.

4.3.2 What Pieces of Land Grant Legislation Were Most Effective?

The current land grant college system is the result of several pieces of legislation, from the 1862 Morrill Act to the most recent farm bill, each of which affected the local innovation ecosystem in different ways. To speak of "the effect" of land grant colleges is therefore to obscure many distinctions that may be important for policy makers. As a first pass at understanding which pieces of legislation had the largest local effect, I repeat the basic differences-in-differences analysis from above but define multiple "postperiod" dummy variables that are equal to one during time periods that denote given legislative epochs. I examine the difference between land grant college counties and runner-up counties following the initial establishment of land grant colleges under the Morrill Act of 1862, the establishment of agricultural experiment stations following the Hatch Act of 1887, and the post–World War II era in which the federal government became much more directly involved in research funding, exemplified by the 1946 Research and Marketing Act.[19] Each of these dates marks a commonly recognized turning point in the funding of higher education, particularly in relation to agricultural research. Numerous studies highlight the pioneering role of the 1862

19. Many other important pieces of legislation could be studied as well, such as the Second Morrill Act of 1890, which established additional land grant colleges, especially for African Americans; the 1906 Adams Act, which provided additional federal funding for scientific research; the 1925 Purnell Act, which provided federal funding for applied research to aid the local agricultural sector; or the 1935 Bankhead-Jones Act, which introduced formula funding and federal and state matching grants for basic agricultural research. Alston and Pardey (1996) provide a useful summary of major legislation related to agricultural research. In additional untabulated analysis, I consider the effects of these other pieces of legislation as well. Unfortunately, many of the acts occurred within a decade or two of one another, making it extremely difficult to separate the effects of particular laws. I therefore focus on what I consider the most important changes in legislation, with the caveat that additional research is needed to conclusively determine the effects of each policy.

Morrill Act in establishing institutions dedicated to agricultural education and research, including several full-length histories (Edmond 1978; Cross 1999, 77–94; Geiger and Sorber 2013; Sorber 2018). A sizable literature also examines the effects of the 1887 Hatch Act, which established state agricultural experiment stations and provided federal funding to conduct research at those stations, marking the beginning of direct federal funding of agricultural research activities (Kerr 1987; Ferleger 1990; Hillison 1996; Kantor and Whalley 2019). The 1946 Research and Marketing Act, which dramatically increased federal spending on state agricultural experiment stations and reorganized the administration of federal agricultural research support, has been the least examined by historians of agriculture or education, although it has not been completely ignored (Bowers 1982; Alston and Pardey 1996). More broadly, the 1946 act exemplifies the federal government's changing approach in the postwar world, with the end of World War II widely recognized as a watershed moment in the federal government's support for university research (Geiger 1993; Rosenberg and Nelson 1994; Mowery and Rosenberg 1998; Mowery and Sampat 2001).

I estimate the following model:

$$(3)\quad Y_{it} = \beta_1 \text{LandGrantCounty}_i \times \text{PostMorrillAct}_{it} + \beta_2 \text{LandGrantCounty}_i \times \text{PostHatchAct}_{it} + \beta_3 \text{LandGrantCounty}_i \times \text{PostWorldWarII}_{it} + \text{County}_i + \text{Year}_t + \varepsilon_{it},$$

where PostMorrilAct equals one for $1862 \leq t < 1887$, PostHatchAct equals one for $1887 \leq t < 1946$, and PostWorldWarII equals one for $1946 \leq t$.[20] I focus on the first cohort of land grant colleges, established between 1862 and 1870, to see how a constant set of colleges changes over the life cycle.

I present results in table 4.6. When splitting up the patenting results into four time periods (the period before the 1862 Morrill Act, which is the base time, and the time periods corresponding to each of the three interaction terms), individual coefficients are typically not statistically significant. It appears that the college counties only begin to see larger levels of patenting relative to the runners-up after the passage of the Hatch Act, with an even larger increase observed after World War II. Agricultural patenting, however, exhibits a different pattern, with the increase in the level of agricultural patents growing in college counties relative to runners-up immediately following the passage of the Morrill Act, falling to almost zero following the Hatch Act, and finally rebounding after World War II. The fraction of agricultural patents appears to increase in land grant college counties relative to the runners-up after the Morrill and Hatch Acts but decreases after

20. Results are similar when replacing the year fixed effects with the much coarser time period dummies for PostMorrilAct, PostHatchAct, and PostWorldWarII.

Table 4.6 **Comparing land grant college counties to runner-up counties following several pieces of legislation**

	log(patents + 1)	log(ag. patents + 1)	Frac. ag. patents	log(total pop.)	log(frac. urban)
	A. Innovation and population outcomes				
College * Post–Morrill Act	−0.0165	0.108	0.0643	−0.0151	−0.0202
	(0.255)	(0.152)	(0.0453)	(0.210)	(0.0330)
College * Post–Hatch Act	0.466	0.0238	0.0389	0.112	0.0182
	(0.340)	(0.0914)	(0.0305)	(0.289)	(0.0420)
College * Post–World War II	0.646	0.179	−0.00594	0.538**	0.0911
	(0.332)	(0.0914)	(0.0100)	(0.156)	(0.0587)
Num. counties × years	4,451	4,451	3,526	4,378	3,538
Adj. R^2	0.747	0.304	0.0582	0.846	0.744

	log(ag. yields)	log(value agricultural output + 1)	log(value crops + 1)	log(value livestock products + 1)
	B. Agricultural outcomes			
College * Post–Morrill Act	−0.0765	−0.128	−0.107	0.228
	(0.0810)	(0.302)	(0.264)	(0.423)
College * Post–Hatch Act	−0.0280	−0.0222	−0.0692	−0.161
	(0.137)	(0.357)	(0.376)	(0.324)
College * Post–World War II	0.0459	0.0971	0.106	0.0909
	(0.0800)	(0.261)	(0.426)	(0.238)
Num. counties × years	4,188	4,398	4,398	4,398
Adj. R^2	0.951	0.947	0.973	0.950

Standard errors in parentheses. * $p < 0.05$, ** $p < 0.01$, *** $p < 0.001$.

Note: Differences-in-differences regression results comparing college counties to runner-up counties before and after several major land grant–related pieces of legislation for the cohort of land grant colleges established between 1860 and 1870. Panel (a) uses innovation and population outcomes as the dependent variables. Panel (b) uses agricultural yield and output as the dependent variables. All regressions include county and year fixed effects. Standard errors are clustered at the county level.

World War II, although the post–World War II magnitude is small.[21] Population and urbanization exhibit increases in college counties relative to the runners-up that are large in magnitude following World War II: total population increases by a statistically significant 54 log points, with urbanization increases by 9 log points. Total population shows a sizable 11 log point increase following the Hatch Act as well. For agricultural yield, agricultural

21. Because the data on the introduction of new wheat varieties are from a 1922 report (Clark, Martin, and Ball 1922), no post–World War II observations are available, and so I do not examine that outcome variable in table 4.6.

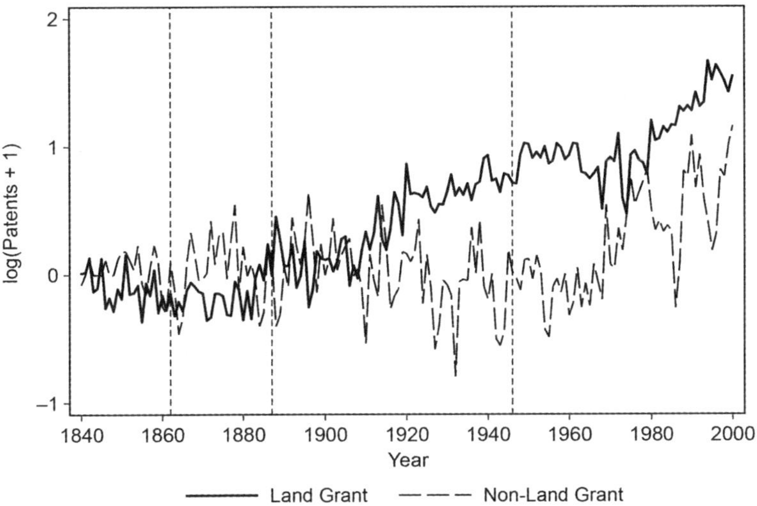

Fig. 4.3 Difference between college and runner-up counties for land grant colleges and non–land grant colleges, calendar time

Note: Plot of the difference in logged patenting between college and runner-up counties for land grant colleges (solid lines) and non–land grant colleges (dashed lines) established between 1860 and 1870. The *x* axis shows calendar years.

output, and crop output, the land grant college counties see a decrease relative to the runner-up counties following the Morrill and Hatch Acts before seeing increases after World War II, although most of these coefficients are fairly small in magnitude, with magnitudes between 2 and 13 log points. Livestock products actually exhibit the largest increase in college counties relative to the runners-up in the years following the Morrill Act, making it difficult to tell a consistent story about the role of each piece of legislation on local agricultural outcomes.

While suggestive, interpreting the results in table 4.6 is difficult. New colleges began as very small institutions that then grew over time, raising the possibility that larger differences between the college and runner-up counties after 1887 or 1946 are driven by the "natural" growth of these colleges rather than by specific policies. To attempt to account for this, I compare the effect of the 1862–70 land grant colleges to the effect of other types of colleges that were established between 1860 and 1870.

Figure 4.3 shows the difference in patenting between college and runner-up counties for this cohort of colleges, where calendar years are plotted on the *x*-axis and the passage of the Morrill, Hatch, and Research and Marketing Acts are indicated. The land grant college counties see sizable increases in the number of patents relative to the runner-up counties beginning in the early 1900s, while a similar takeoff for the non–land grant college counties does not begin until about 1960.[22] To formalize these findings, I estimate the following:

$$
\begin{aligned}
(4)\quad Y_{it} = {} & \beta_1 \text{CollegeCounty}_i \times \text{PostMorrillAct}_{it} \times \text{LandGrant}_i \\
& + \beta_2 \text{CollegeCounty}_i \times \text{PostHatchAct}_{it} \times \text{LandGrant}_i \\
& + \beta_3 \text{CollegeCounty}_i \times \text{PostWorldWarII}_{it} \times \text{LandGrant}_i \\
& + \beta_4 \text{CollegeCounty}_i \times \text{PostMorrillAct}_{it} + \beta_5 \text{CollegeCounty}_i \\
& \times \text{PostHatchAct}_{it} + \beta_6 \text{CollegeCounty}_i \times \text{PostWorldWarII}_{it} \\
& + \beta_7 \text{LandGrant}_i \times \text{PostMorrillAct}_{it} + \beta_8 \text{LandGrant}_i \\
& \times \text{PostHatchAct}_{it} + \beta_9 \text{LandGrant}_i \times \text{PostWorldWarII}_{it} \\
& + \text{County}_i + \text{Year}_t + \varepsilon_{it}.
\end{aligned}
$$

The triple interaction terms $\beta_1 - \beta_3$ show the effect of establishing a land grant college relative to the effect of establishing other types of colleges in each time period. The interaction terms $\beta_4 - \beta_6$ show the average effect of establishing non–land grant colleges in each time period, while the interaction terms $\beta_7 - \beta_9$ show the difference between all counties under consideration to receive a land grant college and all counties under consideration for other types of colleges in each time period. The assumption needed to identify the triple interactions terms of interest is that, without the research-related legislation, land grant and non–land grant colleges of the same age would have similar effects on the local economy at every point in time.

Results are presented in table 4.7. For readability, I only present coefficient estimates for the triple interactions terms, $\beta_1 - \beta_3$; full results are available upon request. All coefficients of interest are—again not surprisingly—not statistically significant, but many are large in magnitude. After the Morrill Act, land grant colleges have roughly 15 log points less of an increase in local patenting than do the non–land grant colleges. This reverses after the Hatch Act, with land grant colleges increasing local patenting relative to their runner-up counties by 46 log points more than the non–land grant colleges after the Hatch Act and 37 log points more after World War II. Land grant colleges have larger increases in the level of agricultural patenting than do the non–land grant patents for all three periods, although in all periods, the land grant colleges see a decline in the share of agricultural patents relative to the non–land grant colleges, with the largest decline in the share of 7 log points occurring after the passage of the Hatch Act.

22. The differences in the relative dynamics of patenting between figures 4.3 and 4.2 is due to the fact that the figures are plotting patenting for a different sample of colleges, with figure 4.3 containing only the schools established between 1860 and 1870.

Table 4.7 **Comparing the land grant to non–land grant colleges following several pieces of legislation**

	log(patents + 1)	log(ag. patents + 1)	Frac. ag. patents	log(total pop.)	log(frac. urban)
	A. Innovation and population outcomes				
College * Post–Morrill Act * LandGrant	−0.149	0.149	−0.0372	−0.276	0.00831
	(0.347)	(0.146)	(0.0797)	(0.213)	(0.0398)
College * Post–Hatch Act * LandGrant	0.456	0.0250	−0.0696	−0.148	0.0193
	(0.441)	(0.0932)	(0.0759)	(0.317)	(0.0662)
College * Post–World War II * LandGrant	0.365	0.165	−0.0131	0.246	0.0460
	(0.407)	(0.107)	(0.0524)	(0.218)	(0.0846)
Num. counties × years	7,248	7,248	5,253	7,227	5,817
Adj. R^2	0.750	0.289	0.0454	0.868	0.771

	log(Ag. Yields)	log(Value Agricultural Output + 1)	log(Value Crops + 1)	log(Value Livestock Products + 1)
	B. Agricultural outcomes			
College * Post–Morrill Act * LandGrant	−0.0598	−0.0179	−0.152	0.179
	(0.162)	(0.481)	(0.398)	(0.692)
College * Post–Hatch Act * LandGrant	−0.0199	0.153	0.126	0.0783
	(0.144)	(0.480)	(0.505)	(0.462)
College * Post–World War II * LandGrant	0.0941	0.302	0.388	0.145
	(0.0953)	(0.344)	(0.679)	(0.289)
Num. counties × years	6,947	7,267	7,267	7,267
Adj. R^2	0.956	0.957	0.976	0.956

Standard errors in parentheses. * $p < 0.05$, ** $p < 0.01$, *** $p < 0.001$.

Note: Triple difference regression results comparing land grant college counties to runner-up counties before and after several major land grant–related pieces of legislation for the cohort of land grant and non–land grant colleges established between 1860 and 1870. Panel (a) uses innovation and population outcomes as the dependent variables. Panel (b) uses agricultural yield and output as the dependent variables. All regressions include county and year fixed effects. Standard errors are clustered at the county level.

The land grant colleges see less of an increase in population after the Morrill and Hatch Acts than do the non–land grant colleges, although following World War II, the land grant colleges have a roughly 25 log points larger increase in population than do the non–land grant colleges. In all three periods, the land grant colleges have larger increases in urbanization (or, at least, less of a decrease), although the magnitudes are very small until after World War II. Land grant colleges have a larger increase in agricultural yield only after World War II, although they have an increase in agricultural output and crop output following the Hatch Act as well and an increase in the value of livestock products sold in all three periods. If anything, land grant colleges see a decline in agricultural yield, agricultural output, and crop output relative to the non–land grant colleges in the initial decades following the passage of the Morrill Act.

These results should be interpreted with caution for several reasons. First, every college that received funding through the Hatch Act also received funding through the earlier Morrill Act; thus the coefficients on Hatch Act funding should be interpreted as the effect of Hatch Act funding conditional on also receiving Morrill Act funding, and the coefficients on post–World War II funding should be interpreted similarly. This point is important to the extent that Hatch Act funding complemented rather than substituted Morrill Act funding, building on institutions and programs established under the earlier law. Second and related, because all the land grant colleges receive funding under all three acts, it is impossible to identify the effects of the Hatch Act from those of the Morrill Act that take several decades to manifest. This is less of a concern when interpreting the coefficients on the Research and Marketing Act, which went into effect almost six decades after the Hatch Act. Finally, none of the triple interaction terms are statistically different from zero; while not surprising given the sample sizes involved, one should refrain from drawing dramatic conclusions from these results.

Despite these caveats, facilitating comparisons of different types of institutions over distinct epochs of federal involvement in agricultural research opens the door to many interesting lines of study. Changes that occur in the postwar period are particularly interesting, because while legislation such as the 1946 Research and Marketing Act specifically targeted agricultural research that was largely conducted at land grant colleges, postwar federal involvement in science and research occurred in nearly all sectors, not merely agriculture.[23] The fact that land grant colleges had a long-established history of supporting applied research may have made land grant colleges a particularly attractive destination of federal funding in the postwar era; I leave a deeper exploration of this issue to future work.

4.4 Conclusion

In this chapter, I provide detailed descriptions of the processes through which states decided where to locate their land grant colleges. Serendipity frequently played a role in determining college location, and I exploit this fact to identify runner-up sites that would have received land grant colleges but for as-good-as-random reasons.

Using these runner-up sites as counterfactuals for locations that receive a land grant college, I show that local agricultural innovation, measured both by patents and new crop varieties, increases in college counties relative to the runners-up after establishing a land grant college. While land grant

23. One may worry that only a few federal institutions dominated postwar federal funding and that these institutions are missing from my sample. O'Mara (2005), for example, documents how skewed federal funding was across institutions. While MIT and Stanford are not in my sample, Georgia Tech (which massively increased its share of federal funding in the 1960s and 1970s) is included as a non–land grant college.

colleges see a sizable increase in innovation, they have small and imprecisely estimated improvements in agricultural performance relative to the runner-up counties. These results lend themselves to several interpretations. One interpretation is that innovations developed at land grant colleges diffuse effectively, but it could also be the case that land grant college innovations have limited relevance to farmers working within the same state. Additional research is needed to determine how the diffusion process for land grant innovations operates. Kantor and Whalley (2019) provide a promising first step in this direction, focusing on the role of geographic proximity and communications technologies in explaining the diffusion from land grant colleges, but much work remains to be done.

More work is also needed to understand exactly what types of policies led to the success of the land grant program and which of these policies can be replicated in other contexts or with other types of institutions. In this chapter, I present suggestive evidence that the Hatch Act and post–World War II federal funding, both of which provided direct federal support for agricultural research, were particularly effective in promoting local invention. Limited variation in the implementation of similar large-scale policies makes these types of questions difficult to answer today. While the historical evidence presented in this chapter is not conclusive, my hope is that the data and methodology presented here will prove to be of continuing utility in addressing important questions for agricultural innovation policy.

Appendix

Table 4.A1 List of non–land grant college experiments

	College	County	State	Runner-up counties	Year established
1	University of Mississippi	Lafayette	Mississippi	Rankin; Attala; Harrison; Montgomery; Winston; Monroe	1841
2	Eastern Michigan University	Washtenaw	Michigan	Jackson	1849
3	College of New Jersey	Mercer	New Jersey	Middlesex; Essex; Burlington	1855
4	University of South Dakota	Clay	South Dakota	Bon Homme; Yankton	1862
5	University of Kansas	Douglas	Kansas	Shawnee	1863
6	Lincoln College (IL)	Logan	Illinois	Edgar; Warrick; Macon	1864
7	Southern Illinois University	Jackson	Illinois	Perry; Clinton; Marion; Washington; Jefferson	1869
8	Mercer University	Bibb	Georgia	Spalding	1870
9	Missouri University of Science and Technology	Phelps	Missouri	Iron	1870
10	University of Oregon	Lane	Oregon	Washington; Linn; Polk	1872
11	University of Colorado	Boulder	Colorado	Fremont	1874
12	University of Texas Austin	Travis	Texas	Smith	1881
13	University of Texas Medical Branch	Galveston	Texas	Harris	1881
14	University of North Dakota	Grand Forks	North Dakota	Burleigh; Stutsman	1883
15	Arizona State University	Maricopa	Arizona	Pinal	1885
16	Georgia Institute of Technology	Fulton	Georgia	Clarke; Greene; Baldwin; Bibb	1886
17	Kentucky State University	Franklin	Kentucky	Boyle; Warren; Daviess; Christian; Fayette	1886
18	New Mexico Tech	Socorro	New Mexico	San Miguel	1889
19	University of New Mexico	Bernalillo	New Mexico	San Miguel	1889
20	Alabama Agricultural and Mechanical University	Madison	Alabama	Montgomery	1891

21	North Carolina A and T University	Guilford	North Carolina	Durham; New Hanover; Alamance; Forsyth	1892
22	Northern Illinois University	DeKalb	Illinois	Winnebago	1895
23	Western Illinois University	McDonough	Illinois	Adams; Hancock; Warren; Schuyler; Mercer	1899
24	University of Nebraska at Kearney	Buffalo	Nebraska	Custer; Valley	1903
25	Western Michigan University	Kalamazoo	Michigan	Barry; Allegan	1903
26	Georgia Southern College	Bulloch	Georgia	Tattnall; Emanuel	1906
27	East Carolina University	Pitt	North Carolina	Lenoir; Beaufort; Edgecombe	1907
28	Middle Tennessee State University	Rutherford	Tennessee	Montgomery	1909
29	Western State Colorado University	Gunnison	Colorado	Garfield; Mesa	1909
30	Arkansas Tech University	Pope	Arkansas	Sebastian; Conway; Franklin	1910
31	Bowling Green State University	Wood	Ohio	Henry; Van Wert; Sandusky	1910
32	Kent State University	Portage	Ohio	Trumbull	1910
33	Southern Arkansas University	Columbia	Arkansas	Hempstead; Ouachita; Polk	1910
34	Southern Mississippi University	Forrest	Mississippi	Jones; Hinds	1910
35	Texas Christian University	Tarrant	Texas	Dallas	1910
36	Southern Methodist University	Dallas	Texas	Tarrant	1911
37	High Point University	Guilford	North Carolina	Alamance	1921
38	Texas Tech	Lubbock	Texas	Scurry; Nolan	1923
39	Maine Maritime Academy	Hancock	Maine	Sagadahoc	1941
40	US Merchant Marine Academy	Nassau	New York	Bristol	1941
41	US Air Force Academy	El Paso	Colorado	Walworth; Madison	1954

Note: List of non–land grant college experiments in the sample, along with the winning county and state, the runner-up counties, and the year in which the site selection decision took place.

Table 4.A2 **Summary statistics of non–land grant college experiments**

	N	Mean	SD	Min	Median	Max
# Runner-up counties	41	2.12	1.38	1.00	2.00	6.00
Distance to college	87	139.57	202.22	30.61	84.77	1,413.28
Year established	41	1893.51	24.99	1841	1892	1954

Note: Number of runner-up counties, average distance from the runner-up counties to the college site, and experiment year for the non–land grant college experiments in the sample.

References

Alston, J. M. 2002. "Spillovers." *Australian Journal of Agricultural and Resource Economics* 46 (3): 315–46.

Alston, J. M., and P. G. Pardey. 1996. *Making Science Pay: The Economics of Agricultural R&D Policy*. Washington, DC: American Enterprise Institute Press.

Andrews, M. J. 2021a. "Historical Appendix: How Do Institutions of Higher Education Affect Local Invention? Evidence from the Establishment of U.S. Colleges." Unpublished manuscript, University of Maryland, Baltimore County, accessed February 14, 2021.

———. 2021b. "How Do Institutions of Higher Education Affect Local Invention? Evidence from the Establishment of U.S. Colleges." Unpublished manuscript, University of Maryland, Baltimore County, accessed February 14, 2021.

Berkes, E. 2018. "Comprehensive Universe of U.S. Patents (CUSP): Data and Facts." Unpublished manuscript, Ohio State University, accessed February 14, 2021.

Bishop, M. 1962. *A History of Cornell*. Drawings by Alison Mason Kingsbury. Ithaca, NY: Cornell University Press.

Bowers, D. E. 1982. "The Research and Marketing Act of 1946 and Its Effects on Agricultural Marketing Research." *Agricultural History* 56 (1): 249–63.

Burnes, B. 2014. *Missou175: The Remarkable Story of Missouri's Flagship University from 1839 to 2014*. Kansas City, MO: Rockhill Books.

Clark, J. A., J. H. Martin, and C. R. Ball. 1922. *Classification of American Wheat Varieties*. US Department of Agrigulture Bulletin no. 1074, November 8, 1922. Revised August 1923. Washington, DC: Government Printing Office.

Cline, P. 1983. *Mountain Campus: The Story of Northern Arizona University*. Flagstaff, AZ: Northland Press.

Cochrane, W. W. 1979. *The Development of American Agriculture: A Historical Analysis*. Minneapolis: University of Minnesota Press.

Cross, C. F. 1999. *Justin Smith Morrill: Father of the Land-Grant Colleges*. East Lansing: Michigan State University Press.

Dethloff, H. C. 1975. *A Centennial History of Texas A&M University, 1876–1976*. Vol. 1. College Station: Texas A&M University Press.

Doten, S. B. 1924. *An Illustrated History of the University of Nevada*. Reno: University of Nevada Press.

Dunaway, W. F. 1946. *History of the Pennsylvania State College*. Lancaster, PA: Lancaster Press.

Edmond, J. B. 1978. *Magnificent Charter: The Origin and Role of the Morrill Land-Grant Colleges and Universities*. New York: Exposition Press.

Ferleger, L. 1990. "Uplifting American Agriculture: Experiment Station Scientists and the Office of Experiment Stations in the Early Years after the Hatch Act." *Agricultural History* 64 (2): 5–23.

Ferrier, W. W. 1930. *Origin and Development of the University of California*. Berkeley, CA: Sather Gate Book Shop.

Fleming, W. L. 1936. *Louisiana State University: 1860–1896*. Baton Rouge: Louisiana State University Press.

Geiger, L. G. 1958. *University of the Northern Plains: A History of the University of North Dakota, 1883–1958*. Grand Forks: University of North Dakota Press.

Geiger, R. L. 1993. *Research and Relevant Knowledge: American Research Universities since World War II*. Oxford: Oxford University Press.

Geiger, R. L., and N. M. Sorber, eds. 2013. *The Land-Grant Colleges and the Reshaping of American Higher Education*. New Brunswick, NJ: Transaction.

Greenstone, M., R. Hornbeck, and E. Moretti. 2010. "Identifying Agglomeration Spillovers: Evidence from Winners and Losers of Large Plant Openings." *Journal of Political Economy* 118 (3): 536–98.

Haines, M., P. Fishback, and P. Rhode. 2018. "United States Agricultural Data, 1840–2012." ICPSR 35206. Accessed May 12, 2019. https://www.icpsr.umich.edu/icpsrweb/ICPSR/studies/35206/publications.

Hall, B. H., A. B. Jaffe, and M. Trajtenberg. 2001. "The NBER Patent Citation Data File: Lessons, Insights and Methodological Tools." NBER Working Paper No. 8498. Cambridge, MA: National Bureau of Economic Research.

Harrington, J., and M. B. Sauter. 2018. "The 25 Most Innovative Cities in US Share Affinity for Technology." *USA Today*, November 14, 2018. https://www.usatoday.com/story/money/2018/11/14/top-cities-entrepreneurs-tech-25-most-innovative-places-us/38330651/.

Hausman, N. 2017. "University Innovation and Local Economic Growth." Unpublished manuscript, Hebrew University, accessed February 14, 2021.

Hayami, Y., and V. W. Ruttan. 1985. *Agricultural Development: An International Perspective*. Baltimore: Johns Hopkins University Press.

Hillison, J. 1996. "The Origins of Agriscience: Or Where Did All That Scientific Agriculture Come From?" *Journal of Agricultural Education* 37 (4): 8–13.

Im, J. 2019. "These Are the 10 Best Places to Live in the US in 2019." CNBC, April 29, 2019. https://www.cnbc.com/2019/04/15/us-news-world-report-best-places-to-live-in-the-us-in-2019.html.

Jaffe, A. B. 1989. "Real Effects of Academic Research." *American Economic Review* 79 (5): 957–70.

Kammen, C. 2003. *Cornell: Glorious to View*. Ithaca, NY: Cornell University Library.

Kantor, S., and A. Whalley. 2014. "Knowledge Spillovers from Research Universities: Evidence from Endowment Value Shocks." *Review of Economics and Statistics* 96 (1): 171–88.

———. 2019. "Research Proximity and Productivity: Long-Term Evidence from Agriculture." *Journal of Political Economy* 127 (2): 819–54.

Kerr, N. A. 1987. *The Legacy: A Centennial History of the State Agricultural Experiment Stations, 1887–1987*. Columbia: Missouri Agricultural Experiment Station, University of Missouri.

Manson, S., J. Schroeder, D. V. Riper, and S. Ruggles. 2018. IPUMS National Historical Geographic Information System: Version 13.0. http://doi.org/10.18128/D050.V13.0.

Marco, A. C., M. Carley, S. Jackson, and A. F. Myers. 2015. "The USPTO Historical Patent Data Files: Two Centuries of Invention." Unpublished, USPTO Economic Working Paper No. 2015–1, Alexandria, VA.

Maredia, M. K., R. Ward, and D. R. Byerlee. 1996. "Econometric Estimation of a Global Spillover Matrix for Wheat Varietal Technology." *Agricultural Economics* 14 (3): 159–73.
Martin, D. D. 1960. *The Lamp in the Desert: The Story of the University of Arizona*. Tucson: University of Arizona Press.
McMath, R. C., Jr., R. H. Bayor, J. E. Brittain, L. Foster, A. W. Giebelhaus, and G. M. Reed. 1985. *Engineering the New South: Georgia Tech, 1885–1985*. Athens: University of Georgia Press.
Montgomery, J. R., S. J. Folmsbee, and L. S. Greene. 1984. *To Foster Knowledge: A History of the University of Tennessee, 1794–1970*. Knoxville: University of Tennessee Press.
Moscona, J. 2019. "Downstream Effects of Research Incentives: Evidence from Agricultural Innovation." Unpublished manuscript, MIT, accessed February 14, 2021.
Moser, P., and P. W. Rhode. 2012. "Did Plant Patents Create the American Rose?" In *The Rate and Direction of Inventive Activity Revisited*, edited by J. Lerner and S. Stern, 413–38. Chicago: University of Chicago Press.
Mowery, D. C., and N. Rosenberg. 1998. *Paths of Innovation: Technological Change in 20th-Century America*. Cambridge: Cambridge University Press.
Mowery, D. C., and B. N. Sampat. 2001. "University Patents and Patent Policy Debates in the USA, 1925–1980." *Industrial and Corporate Change* 10 (3): 781–814.
Olmstead, A. L., and P. W. Rhode. 2008. *Creating Abundance: Biological Innovation and American Agricultural Development*. Cambridge: Cambridge University Press.
O'Mara, M. P. 2005. *Cities of Knowledge: Cold War Science and the Search for the Next Silicon Valley*. Princeton, NJ: Princeton University Press.
Proctor, S., and W. Langley. 1986. *Gator History: A Pictorial History of the University of Florida*. Gainesville, FL: South Star.
Reel, J. V. 2011. *The High Seminary: A History of the Clemson Agricultural College of South Carolina, 1889–1964*. Vol. 1. Clemson, SC: Clemson University Digital Press.
Rees, D., and A. Walsworth. 1989. *The University of Missouri: 150 Years*. Marceline, MO: Walsworth.
Reynolds, J. H., and D. Y. Thomas. 1910. *History of the University of Arkansas*. Fayetteville: University of Arkansas Press.
Roberts, C. N. 1946. *History of the University of Missouri School of Mines and Metallurgy, 1871–1946*. Rolla, MO: Missouri School of Mines and Metallurgy.
Rosenberg, N., and R. R. Nelson. 1994. "American Universities and Technical Advance in Industry." *Research Policy* 23 (3): 323–48.
Ross, E. D. 1958. *The Land-Grant Idea at Iowa State College: A Centennial Trial Balance, 1858–1958*. Ames: Iowa State College Press.
Sansing, D. G. 1999. *The University of Mississippi: A Sesquicentenniel History*. Oxford: University Press of Mississippi.
Scheuring, A. F. 2001. *Abundant Harvest: The History of the University of California, Davis*. Davis, CA: UC Davis History Project.
Smith, D. C. 1979. *The First Century: A History of the University of Maine, 1865–1965*. Orono: University of Maine at Orono Press.
Solberg, W. U. 1968. *The University of Illinois, 1867–1894: An Intellectual and Cultural History*. Urbana: University of Illinois Press.
Sorber, N. M. 2018. *Land-Grant Colleges and Popular Revolt: The Origins of the Morrill Act and the Reform of Higher Education*. Ithaca, NY: Cornell University Press.
Turner, F. H. 1932. "Misconceptions Concerning the Early History of the University of Illinois." *Illinois State Historical Society Transactions* 39:63–90.

Wagoner, J. J. 1970. *Arizona Territory, 1863–1912: A Political History*. Tucson: University of Arizona Press.
Willis, J. F. 2009. *Southern Arkansas University: The Mulerider School's Centennial History, 1909–2009*. Bloomington, IN: Xlibris.
Wright, B. D. 2012. "Grand Missions of Agricultural Innovation." *Research Policy* 41 (10): 1716–28.

Comment Bhaven N. Sampat

4.C1 Background

This chapter examines the effects of land grant universities on local innovation and agricultural output. It is a useful contribution not only to the literature on agricultural innovation but also to the broader literature on returns from publicly funded research.

In previous work, I have been among those who have pointed to the land grant college system as an exemplar of university applied research and dissemination working well. In particular, I have held up the land grant system as a good model of technology and knowledge transfer and as perhaps better at securing social returns from publicly funded research than the current system focused on patenting, licensing, and technology transfer (Mowery et al. 2004).

Reading this chapter led me to rethink this.

4.C2 Summary

As the chapter indicates, a big problem in the economics literature examining the effects of universities on local outcomes is that universities are not randomly located.

Through meticulous (and what seems like very labor intensive but also fun!) historical research, Andrews finds the cases where the location of the land grant university within a state was chosen through an "as good as random" process and focuses empirical analyses on these 29 universities. A big contribution of this chapter is laying out the site choice decision, which points to the importance of politics, personalities, and happenstance

Bhaven N. Sampat is an associate professor in the Department of Health Policy and Management at Columbia University and a research associate of the National Bureau of Economic Research.

For acknowledgments, sources of research support, and disclosure of the author's material financial relationships, if any, please see https://www.nber.org/books-and-chapters/economics-research-and-innovation-agriculture/comment-local-effects-land-grant-colleges-agricultural-innovation-and-output-sampat.

in these decisions. This alone will be useful for future empirical research on the economics of agricultural research.

Next, Andrews tests whether the establishment of a land grant college had effects on local innovation and agricultural output. The identification approach involves comparing outcomes for the land grant counties versus the "good as random" runner-ups.

The chapter finds that measures of agricultural innovation (patents and new crop varieties) increase in these counties. These effects are large and statistically significant. However, when the chapter looks next at measures of agricultural performance (agricultural yields, the value of agricultural output, the value of crop output, the value of livestock produced), there do not appear to be strong local effects.

This leaves us with a puzzle. Why, if innovation increases locally, including agricultural innovation, do we fail to see local productivity effects? As the chapter points out, one potential explanation is that the innovation was not very useful to local farmers. Another is that it was useful, but it diffused broadly throughout the state, including to the runner-up county. These different explanations would obviously lead us to very different assessments of the land grant universities from a national perspective or that of state taxpayers.

Unfortunately, it is not possible to test directly which of the competing explanations is more plausible given the data that are available. As a second best, Andrews does provide some evidence from a catalog of wheat varieties, showing that land grant universities came up with about 30 percent of new wheat varieties, but these accounted for only 10 percent of national acreage. That is, yes, there was wheat innovation, but it did not broadly diffuse. One thing that we do not learn from this exercise is whether these innovations were useful locally (in the land grant counties themselves), but that too likely reflects data constraints.

4.C3 Suggestions

The next steps in this line of research would seem to be to better distinguish between the competing explanations. I have a few other observations and suggestions as well.

First, the finding that only about 10 percent of wheat acreage in home states was from land grants contrasts with previous, more positive assessments of the Morrill Act (Wright 2012), which suggest that three-fourths of wheat acreage by 1920 used wheat varieties that were unavailable when the act passed.

But it also seems possible that the land grant research had an indirect effect on productivity—producing new research techniques and science rather than new varieties themselves. I read the literature on the economic impact of universities as suggesting that actual products from academic research are less important than research techniques and tools (Cohen, Nelson, and Walsh 2002). Indeed, Griliches's famous article on hybrid corn emphasizes that "hybrid corn was the invention of a method of inventing" (Griliches 1957, 502). My own research in a very different context, drug development over the 1988–2005 period, suggests that public sector labs account directly for about 10 percent of drugs but may enable two-thirds of marketed drugs (Sampat and Lichtenberg 2011). Cockburn and Henderson (2000) find similar orders of magnitude for drugs. All this would seem to suggest that for land grants, tracking disembodied knowledge flows and indirect effects of the universities may be useful going forward. Similarly, it may be interesting to track where graduates went to assess the role of knowledge "wrapped up in people," to paraphrase Robert Oppenheimer (Zolas et al. 2015).

Another explanation for the puzzle that I flagged above—that land grants helped with local innovation but not output—is that the universities were not sufficiently focused on local demand. There is a gap between the innovation that universities do and what the funders want. This echoes broader critiques of public research in science and technology policy, where concern by policy makers that research is not effectively targeted to demand, or not effectively diffused, has ebbed and flowed over the post–World War II era (Brooks 1996; Geiger 2008). I am not exactly sure how to test this without data on research investments and agricultural needs, but it would seem to me that if this were the case for the land grants (or similarly, if diffusion and "translation" of the research findings were not effective), one would see some qualitative evidence in state-level debates about funding. If the land grant innovation is not that useful after all, where are the disgruntled state taxpayers? This seems trackable through testimony, media articles, or the historical literature. The idea that land grant research was not actually that relevant (or that good) would challenge the prevailing understanding on the political economy of land grants, suggesting they were a model of use-oriented research and active dissemination that worked—and worked because they were responsive to local taxpayers. This is precisely why it is important to do.

Though there is a lot more to be done, this chapter, and the author's companion work, represent an excellent start on what will be a very important line of research that will contribute not only to assessment of the land grants and the Morrill Act but also to our understanding of the economics of innovation and diffusion and potentially the political economy of innovation policy. I appreciate the opportunity to comment on it in its germinal stages.

References

Brooks, H. 1996. "The Evolution of US Science Policy." In *Technology, R&D, and the Economy*, edited by Bruce L. R. Smith, Claude E. Barfield, and Paul Dufour, 15–48. Washington, DC: Brookings Institution.

Cockburn, I. M., and R. M. Henderson. 2000. "Publicly Funded Science and the Productivity of the Pharmaceutical Industry." *Innovation Policy and the Economy* 1:1–34.
Cohen, W. M., R. R. Nelson, and J. P. Walsh. 2002. "Links and Impacts: The Influence of Public Research on Industrial R&D." *Management Science* 48 (1): 1–23.
Geiger, R. L. 2008. *Research and Relevant Knowledge: American Research Universities since World War II*. Piscataway, NJ: Transaction.
Griliches, Z. 1957. "Hybrid Corn: An Exploration in the Economics of Technological Change." *Econometrica* 25 (4): 501–22.
Mowery, D. C., R. R. Nelson, B. N. Sampat, and A. A. Ziedonis. 2004. *Ivory Tower and Industrial Innovation: University-Industry Technology Transfer before and after the Bayh-Dole Act*. Palo Alto, CA: Stanford University Press.
Sampat, B. N., and F. R. Lichtenberg. 2011. "What Are the Respective Roles of the Public and Private Sectors in Pharmaceutical Innovation?" *Health Affairs* 30 (2): 332–39.
Wright, B. D. 2012. "Grand Missions of Agricultural Innovation." *Research Policy* 41 (10): 1716–28.
Zolas, N., N. Goldschlag, R. Jarmin, P. Stephan, J. Owen-Smith, R. F. Rosen, B. M. Allen, B. A. Weinberg, and J. I. Lane. 2015. "Wrapping It Up in a Person: Examining Employment and Earnings Outcomes for Ph.D. Recipients." *Science* 350 (6266): 1367–71.

5

Academic Engagement, Commercialization, and Scholarship
Empirical Evidence from Agricultural and Life Scientists at US Land Grant Universities

Bradford Barham, Jeremy Foltz, and Ana Paula Melo

5.1 Introduction

Research on the factors shaping university-industry relations (UIR) has exploded in recent decades, as reflected by the hundreds of recent articles published on this topic.[1] At the heart of this take-off was the push by universities worldwide to pursue opportunities to commercialize intellectual property rights. Arguably, the 1981 passage of the Bayh-Dole Act put US public research universities at the forefront of this global expansion. It extended the intellectual property rights of American universities and their researchers to commercialize innovations and discoveries associated with federally sponsored research (Henderson, Jaffe, and Trajtenberg 1998; Grimaldi et al. 2011; Sampat 2006; Thursby and Thursby 2011). European and universities elsewhere followed suit to varying degrees. In the process, UIR around the globe expanded traditional scholarship models of publish-

Bradford Barham is a professor of agricultural and applied economics at the University of Wisconsin-Madison.

Jeremy Foltz is a professor of agricultural and applied economics at the University of Wisconsin-Madison.

Ana Paula Melo is a PhD candidate in agricultural and applied economics at the University of Wisconsin-Madison.

Thanks to Petra Moser and participants at the NBER "Economics of Research and Innovation in Agriculture" conference for comments, Jordan Van Rijn and Josh Alfonso for data work, and the US Department of Agriculture (USDA) Agriculture and Food Research Initiative (AFRI) and USDA-Hatch grants for funding. For acknowledgments, sources of research support, and disclosure of the authors' material financial relationships, if any, please see https://www.nber.org/books-and-chapters/economics-research-and-innovation-agriculture/academic-engagement-commercialization-and-scholarship-empirical-evidence-agricultural-and-life.

1. See, for example, Agrawal (2001); Djokovic and Souitaris (2008); Geuna and Muscio (2009); Perkmann et al. (2013); and Sengupta and Ray (2017).

ing and training students into directly engaging with industry and entering commercial domains via patents, start-ups, and other forms of corporate-university alliances.

Our study sheds light on the ground level of UIR at leading US land grant universities (LGUs) by examining the activities, attitudes, and research choices of individual agricultural and life science (ALS) faculty. At LGUs, faculty engagement with industry dates back to the end of the 19th century based on an explicit emphasis on practical agricultural and engineering sciences, formal extension appointments for faculty, and ongoing outreach with farms and firms to improve their performance. The recent salience of UIR activities to US LGUs stems from the financial stress they faced over the past three decades due to declines in state and federal support (Just and Huffman 2009; Ehrenberg 2012; Hoag 2005).[2] Hence most US LGUs pursued academic commercialization as a potential mechanism to generate royalties and start-up revenue streams (Thursby and Thursby 2011).

Our chapter exploits rich, unique, and representative individual-level cross-sectional and panel survey data gathered in 2005 and 2015 from ALS faculty from all 52 of the original 1863 US LGUs. We explore the prevalence, intensity, and importance of US land grant faculty's work with industry as compared to traditional scholarship activities. We also examine how faculty attitudes toward research choices shape their participation in UIR and how they combine UIR with traditional scholarship activities. We divide university industry relations into two types.[3] One is academic engagement (AE), defined as faculty participation in sponsored and collaborative research, contract research, consulting, and informal relationships with private firms and institutions. Academic commercialization (AC) is the other, defined as faculty participation in private intellectual property creation (via invention disclosures, patents, and licensing) and entrepreneurship (e.g., start-ups). These definitions are used in other recent articles that examine UIR among university faculty in Europe (D'Este and Perkmann 2011; Perkmann, King, and Pavelin 2011; Tartari, Perkmann, and Salter 2014; Tartari and Salter 2015; Sengupta and Ray 2017).

These apparently contrasting categories are not mutually exclusive types of UIR, with many faculty doing both AE and AC. In our analysis, we contrast faculty who engage in these three categories—engagement (AE), commercialization (AC), and both (AE/AC)—with faculty who are not engaged in any of the three, which we categorize as traditional scholarship (TS).[4]

2. For example, compared to the 2007–8 school year, state spending on higher education, which is a significant portion of LGU budgets nationwide, was down in 2015 by 23 percent, or $2,026 per student (Mitchell, Palacios, and Leachman 2015).

3. We follow the classification adopted in Perkmann et al. (2013).

4. We are cognizant of the long-standing tradition of faculty, especially at LGUs, engaging with industry. Our nomenclature is meant to distinguish between the traditional activities of teaching and research with UIR.

Together, these four categories (AE, AC, AE/AC, and TS) characterize how university faculty engage with industry.

We address four major questions: What is the prevalence and intensity of AE and AC activities among ALS faculty at flagship public research universities across the United States? What role does UIR play in funding faculty research? How do the research and teaching outputs of faculty active in UIR activities compare with those of traditional scholars? And last but certainly not least, how do the UIR activities and attitudes of land grant ALS faculty align with researcher motivations for their choice of research problem? Because UIR activities "tend to be individually driven and pursued on a discretionary basis" (Perkmann et al. 2015, 424), we examine them at the individual faculty level, where we can probe how they meet the values and motivations of faculty. Participation largely depends on the "independent initiative of autonomous, highly skilled" faculty pursuing research and knowledge transfer activities that they value for scientific and/or commercial reasons.

Our work builds on several recent research contributions and is especially motivated by the Perkmann et al. (2013) review on faculty activity in UIR. They identify three major information gaps, which we address directly in this chapter. One is the lack of comparative evidence from US universities regarding faculty engagement in distinct types of UIR activities, since the literature is mostly based on European university data. They also document surprisingly little examination of the two UIR activities (engagement and commercialization) side by side and the factors shaping faculty engagement with them. Although there is a vast body of research on AC and its impacts on faculty scholarship (Agrawal and Henderson 2002; Azoulay, Ding, and Stuart 2007), relatively little research compares and contrasts it with the full set of possible ways for faculty to engage with industry, as we do here. The third is the lack of temporal—including longitudinal—evidence that allows attention to trends over time of innovation in UIR. This is now a relatively mature episode, with the academic commercialization take-off in the United States having occurred by the 1990s and in Europe not long afterward, which warrants study.

Other recent papers help motivate this article. Sengupta and Ray (2017) probe the dynamic relationship between both types of UIR (what they call "knowledge transfer") and traditional research outputs at UK research universities. Using a longitudinal, university-level data set (spanning 2008–14), they find that both AE and AC are positively associated with past research performance. However, consistent with the higher prevalence and intensity of AE relative to AC in UK universities, they also show that only the former has strong positive feedback effects on subsequent research performance via both funding and research scholarship (using both quantity and quality measures). This major finding in the United Kingdom helps set the broad stage for our analysis of UIR and research activities among individual ALS faculty in the major US LGUs.

D'Este and Perkmann (2011) distinguish between two ways in which faculty attitudes toward UIR may shape their participation. In the first, faculty are viewed as academic entrepreneurs who seek to engage in UIR for commercialization reasons, what we refer to as commercial motivation. In the second, faculty are viewed as scientists operating in a strongly institutionalized environment who mainly seek UIR collaborations to advance their research efforts, what we later call intrinsic motivation. We recover these motivations from our data using factor analysis of attitudinal questions and then, in a similar fashion to D'Este and Perkmann, link them to faculty activity choices.

Finally, Perkmann et al. (2013), as well as Sengupta and Ray (2017), highlight the potential importance of university-level infrastructure, research quality, and incentives for promotion and salary increases in shaping faculty engagement with UIR activities. Specifically, the historical experience and current resource base associated with university technology transfer offices can positively shape UIR outcomes. Likewise, universities with higher-quality research performance may be more attractive to industry partners and thus attract UIR. Cutting the other way is the possibility that faculty at the very top universities, especially in some fields, may be less inclined toward applied research and UIR relative to pursuing large public or foundation grants and peer-based collaborations.

Our analysis of the data on LGU faculty answers the four research questions as follows. First, at US LGUs, AE, which includes sponsored research, industry collaborations, and presentations, is far more prevalent and intensively pursued than is AC, which includes patenting, licensing, and start-ups. Several decades into the LGU push toward commercialization, faculty participation appears to have plateaued at much lower levels than their academic engagement. And additional longitudinal evidence shows that AE is the more steadily pursued form of UIR, while AC appears to be more opportunistic, perhaps consistent with the notion that only occasional scientific breakthroughs are worthy of a patent.

Second, commercialization generates very low levels of revenue streams for the operation of LGU faculty research labs. By contrast, funding generated by sponsored research of various sorts (including continued support from commodity organizations) outpaces commercialization revenues by a ratio of about 10:1 and represents for many faculty a substantive portion of their research lab expenditures. Nonetheless, public funding, especially federal funds, continues to be the majority source of faculty lab funding. Thus while AE is far more important as a revenue stream for faculty research activities than is AC, it remains overall a distant second to public funds.

Third, consistent with many previous studies, we find that UIR activities of both types are higher among faculty with higher levels of traditional academic scholarship outputs. Thus UIR and academic scholarship appear to be synergistic, reflecting at an individual level the dynamic feedbacks identified by Sengupta and Ray (2017) in UK data at a university level. This "synergy" finding also implies that concerns about major trade-offs between faculty UIR activity and traditional academic scholarship may be offtrack. Indeed, they appear to overlook a positive dynamic feedback loop that can nourish more of both types of activity over time.

Finally, the regression analysis reveals that individual-, institutional-, and university-level factors all help explain faculty UIR activity. As found elsewhere, attitudes and activity choice align in ways that are consistent with faculty participating in UIR for reasons related to advancing scientific research rather than pursuing commercialization outcomes. The university-level fixed effect results are also intriguing, as they suggest that higher levels of UIR activity are contingent on culture, history, location, and quality of science associated with the overall university (not just individual faculty).

The next section describes the context of colleges of agricultural and life sciences at US LGUs, while section 5.3 introduces the data and explains our methods. Section 5.4 presents the results, while section 5.5 discusses the implications of our findings for UIR in the United States. Section 5.6 concludes.

5.2 US Land Grant Universities

Three major legislative acts frame the long-standing tradition of academic engagement at US LGUs (Fitzgerald et al. 2012). The first is the Morrill Act of 1862, which granted states land to help finance the establishment of public universities. They emphasized agricultural and mechanical arts in support of those two major economic sectors while broadening access to education and training. The second is the Hatch Act of 1887, which provided funding to LGUs to invest in agricultural experimental stations. It recognized the value of increasing public commitment to research that advanced knowledge for both farmers and consumers with respect to production and nutrition/health outcomes. Finally, the Smith-Lever Act of 1914 created the infrastructure for delivering knowledge to society via an extension system. It aimed at both sharing research discoveries with farmers, firms, and consumers and identifying future research issues based on feedback from those and other "stakeholders." Combined, these three acts shaped a long and rich history of AE at US LGUs that featured colleges of agriculture (and later "life sciences") as the cutting edge of UIR activities. Some faculty appointments included explicit attention to "extension" in combination with traditional scholarship: research and instruction duties.

Faculty in US colleges of agricultural and life sciences generally span the breadth of basic and applied sciences reflected across the rest of public research universities. Some departments are filled primarily with basic scientists. This holds especially in "biology" departments, such as genetics, molecular biology, and biochemistry, as well as in "ecology" departments (of various names). There are mostly applied (but some basic) scientists in animal science departments (including specialties in dairy or poultry science), food and nutrition science departments, plant science departments (including agronomy, entomology, horticulture, plant pathology, and soil science), and agricultural or biosystems engineering. Finally, colleges of agricultural and life sciences have social science departments of various names that include economists, sociologists, journalism and communications scholars, and regional planning and community development faculty. While most of these social scientists tend to work on more "applied" questions, there are also some who could be viewed as closer to "basic" in their orientation to pursuing advances on "theory" and "measurement" issues rather than emphasizing applied questions. Thus the fields in US LGUs tend to provide distinctive "institutional" contexts in which to frame the likely connections between faculty and UIR activities.

In the 1990s, as with other universities, AC efforts took off in US LGUs colleges of agricultural and life sciences (Barham, Foltz, and Kim 2002; Foltz, Kim, and Barham 2003; Sampat 2006). Biotechnology patents especially were viewed as a potential source of growth and expansion in both UIR and revenue streams for universities and faculty inventors. A plethora of literature explores this period (Phan and Siegel 2006; Grimaldi et al. 2011), with a primary focus on whether academic activities and the pursuit of open science would be advanced or reduced by the attention to commercialization efforts (Thursby and Thursby 2011). At the "field level," this AC push arguably expanded the potential for higher levels of faculty participation in UIR among more basic scientists who might be able to pursue patents on discoveries more readily than they might seek out sponsored research or active collaboration with industry scientists. Thus it is arguable that AC engagement may be higher among biologists, but the long-standing engagement with AE activities by the more applied scientists could also readily give rise to patenting and commercialization efforts depending on the research topics and discoveries being pursued. These cross-cutting trends make it difficult to envision a clear distinction in terms of AC participation across the natural science fields. On the other hand, social scientists are far less likely to be engaged with patenting and licensing efforts. Most of their "idea" discoveries are likely to be algorithms and statistical or system modeling innovations rather than material ones. As a result, AC participation among social scientists is likely to be lower than other types of science faculty in colleges of agricultural and life sciences.

The rise in US LGU efforts to promote AC coincided with a secular decline in federal and state support for higher education (Ehrenberg 2012). While LGUs were initially able to largely compensate for that decline by raising tuition fees, significant pressures on the research and salary expenditures were experienced especially between 2005 and 2015 (Mitchell, Palacios, and Leachman 2015). During that time period, most LGUs experienced an overall decline in state revenues. Faculty increasingly experienced real declines in salary levels as well as increased pressure to pursue extramural funding of various types—including UIR—to support their labs and salaries (American Academy of Arts and Sciences 2016). Indeed, many colleges of agricultural and life sciences pursued conversions of faculty salary contracts from 12-month to 9-month appointments. Faculty were "incentivized" to pursue the additional 3 months of salary through external sources or "administrative" postings. All of these changes could potentially be viewed as commercial or financial motivations to increase both AE and AC efforts, if in fact they held potential for filling holes in research budgets and faculty summer salary needs.

Two other contextual trends in US LGUs warrant attention here. One is the pressure on research time associated with "changes" in university budgets. As documented in Barham, Foltz, and Prager (2014), US LGU ALS faculty reported declines in "research time" and concomitant increases in time spent on administrative activities. Reducing support staff and increasing faculty reporting efforts is one way in which LGUs dealt with budget cuts and compliance demands. This could have put pressure on faculty to limit UIR as part of the overall pressure on their time, especially research time. The other one, which is "more speculative," is the potential for morale issues associated with this long period of budget pressures and time constraints. It seems likely that these could have either dampened enthusiasm for UIR activities (exhaustion) or increased incentives for faculty to pursue especially commercial links for more personal gain.

5.3 Data, Methods, and Descriptive Statistics

This chapter is based on data collected in surveys of ALS faculty conducted in 2005 and 2015. In each data collection effort, we administered a survey to nearly 3,000 ALS faculty at all of the US LGUs established in 1863.[5] Both surveys had a sample frame that included all tenure-track faculty scientists in ALS departments at these LGUs. We culled faculty names from university web directories to create the cross-sectional sample frame and then randomly selected a sample of scientists who were sent a web-based survey with follow-up paper-mail reminders as in Dillman (2011). In addition to the random samples in both years, we also resampled respondents from the 2005 survey in 2015 in order to have longitudinal data on faculty. The response rate in 2015 was 32 percent based on respondents who answered at least one survey question, with a higher response rate in 2005 of 68 percent.[6]

5. The institutional review board at UW-Madison approved both of these surveys, with the latest approval being #2015–0924.

6. For more about the surveys, see Goldberger et al. (2005) and Barham et al. (2017).

Table 5.1 **Types of UIR and survey items included**

			Survey item description
University-industry relations	Academic engagement	Faculty participation in sponsored and collaborative research, contract research, and information relationships with private firms and institutions.	Research support from private industry Research support from commodity organizations Collaborated with scientists in private industry Coauthored with scientists in private industry Presented to farmers or farm organizations Presented to commodity groups Presented to the private industry Farmers or farm org. helped you identify a research problem Collaborated on a research project with farmers or farm org. Coauthorship on paper or patent with farmers or farm org.
	Academic commercialization	Faculty participation in private intellectual property creation—via invention disclosure, patents, and licensing—and entrepreneurship (e.g., start-ups).	Licensing or patenting revenue # disclosures generated # patent applications generated # patents issued # patents licensed out # products under regulatory review generated # products on the market generated # start-up companies founded

Response rates in 2015 did vary somewhat by discipline, from a high of 42 percent among plant scientists (the largest discipline represented) to only 28 percent among agricultural engineering scientists (the smallest discipline). We accept the null hypothesis of no response rate bias (see Barham et al. 2017) with respect to the following observed characteristics: field, gender, faculty size of the agricultural college, total university research funding, or total full-time university student enrollment. In appendix A, we report further sample restrictions. Our final sample for analysis, from the random sample data collection, covers 925 scientists in 2005 and 615 in 2015 across all 52 LGUs. We also report results from the longitudinal sample of 244 scientists surveyed in both years.

Table 5.1 details the set of questions with respect to faculty UIR activities in our data. AE activities span a similar range described in the aforementioned studies in the United Kingdom. They cover collaborations, sponsored research by industry (and commodity organizations), presentations to industry or farmers, and research problem identification. Likewise, AC activities span invention disclosures, patenting, licensing, product development, and start-ups.

We use these UIR-related items in the data to construct categorical variables of AE and AC participation measures as well as ones that identify when individuals do both. We classify individuals not fitting in any UIR category as traditional scholars. The participation measure is "liberal" in the sense that participating in any of the AE or AC activities identifies an individual with that category. We use these categorical variables to describe trends in UIR participation on the "extensive" margin.

In addition to individual participation in UIR, the subsequent analysis also focuses on other faculty research activities. We mostly focus on publishing articles, training graduate students, and receiving research funding. Those research activities are incorporated into the comparisons of faculty across UIR categories in order to help identify the potential for synergies or trade-offs between UIR and traditional scholarship outcomes. Similarly, we use data on total research grant revenues and different sources of revenue, such as federal, state, industry, commodity groups, foundations, and licensing revenues.

Two other important sets of measures from the survey warrant description here. First, in both 2005 and 2015 surveys, respondents were asked about what motivated them to pursue a certain research topic in the last five years. They are generally oriented toward "intrinsic" motivations, such as "scientific curiosity" or "potential contribution to scientific theory," or extrinsic ones, such as "potential marketability" or "potential to patent and license the discovery." The full set of 14 questions are shared in table 5.8. The items are arranged in a five-point Likert scale, with a score of 1 being "Not at all" and a score of 5 being "Extremely." Responses to these questions are examined using factor analysis in order to uncover latent factors that might shape faculty research choice. We interpret the estimated factor loadings to identify the subset of items with internal consistency, which we classified as two factors: intrinsic and extrinsic motivations to pursue research. We calculate indexes for each measure of intrinsic and extrinsic motivations. Indexes are calculated as the response average for the block of items within each factor as reported in table 5.9.

We begin with three broad observations that start to frame US LGUs' participation in UIR activities. They can be gleaned from tables 5.2, 5.3, and 5.4. Table 5.2 provides a comparison for 2005 and 2015 of the prevalence of each of the UIR activities. Table 5.3 provides a comparison over time of faculty participation in the four UIR categories. In table 5.4, we describe participation rates in AE and AC UIR activities by gender, rank, appointment type, and field.

We find that US LGU faculty participation rates in UIR activities are high (table 5.2), averaging 78 percent of faculty participating in any type of UIR (table 5.3). Consistent with other evidence in the literature, AE is far more prevalent than AC, with about 76 percent of LGU faculty pursuing AE as compared to 19 percent in some type of AC in 2015. Moreover, if we isolate on the exclusive AC category in table 5.3, we find that around 2–3 percent of faculty are just doing AC in the two time periods. In other

Table 5.2 AE and AC activity participation rates and counts, 2005 and 2015

	2005		2015		
	Rate	Count	Rate	Count	Δ p.p.
Academic engagement					
Had research support from private industry	0.47	437	0.45	275	−0.03
Had research support from commodity organizations	0.32	294	0.29	177	−0.03
Collaborated with scientists in private industry	0.29	265	0.36	219	0.07**
Coauthored with scientists in private industry	0.13	118	0.15	93	0.02
Presented to farmers or farm organizations	0.42	385	0.38	233	−0.04
Presented to commodity groups	0.32	299	0.31	188	−0.02
Presented to the private industry	0.32	299	0.29	181	−0.03
Had help from farmers or farm organizations to you identify a research problem	0.37	341	0.38	233	0.01
Collaborated on a research project with farmers or farm organizations	0.27	253	0.31	192	0.04
Coauthorship on paper or patent with farmers or farm organizations	0.03	30	0.03	18	−0.00
Academic commercialization					
Received any royalties income from patent (past five years)	0.04	39	0.05	31	0.01
Had licensing or patenting revenue returned to your research lab (last year)	0.02	23	0.04	23	0.01
Number of disclosures generated	0.16	144	0.13	81	−0.02
Number of patent applications generated	0.16	146	0.11	68	−0.05**
Number of patents issued	0.10	88	0.06	39	−0.03*
Number of patents licensed out	0.04	40	0.04	22	−0.01
Number of products under regulatory review generated	0.02	20	0.01	9	−0.01
Number of products on the market generated	0.07	67	0.05	29	−0.03*
Number of start-up companies founded	0.04	35	0.03	17	−0.01

Table 5.3 Faculty participation rates in UIR, 2005 and 2015

	2005	2015	Diff.
Academic engagement (AE)	0.75	0.76	0.01
Academic commercialization (AC)	0.26	0.19	–0.07**
Mutually exclusive measures			
Academic engagement (AE)—exclusively	0.53	0.60	0.07*
AE and AC	0.22	0.17	–0.05**
Academic commercialization (AC)—exclusively	0.03	0.02	–0.01
Traditional scholarship	0.22	0.22	0.00

words, the vast majority of faculty engaged in AC activities are also active in AE. The proportion of faculty that are not engaged in UIR, the TS category, is greater than the total proportion active in AC. Thus academic commercialization is the least prevalent in the mix of faculty engagement types examined here.

UIR participation declined somewhat between 2005 and 2015. Declines

Table 5.4 **Individual characteristics of UIR categories, 2005 and 2015**

	2005				2015			
	AE	AE/AC	AC	TS	AE	AE/AC	AC	TS
Male	0.54	0.23	0.03	0.20	0.58	0.19	0.02	0.21
Female	0.50	0.17	0.04	0.29	0.63	0.11	0.02	0.24
Rank								
Professor	0.50	0.25	0.03	0.22	0.57	0.21	0.02	0.20
Associate professor	0.59	0.20	0.03	0.18	0.64	0.09	0.01	0.25
Assistant professor	0.54	0.17	0.04	0.24	0.61	0.14	0.03	0.21
Fields								
Ag engineering	0.54	0.33	0.04	0.09	0.58	0.29	0.03	0.10
Animal science	0.59	0.33	0.03	0.05	0.61	0.24	0.02	0.13
Biology	0.19	0.24	0.12	0.45	0.34	0.18	0.11	0.38
Plant science	0.59	0.28	0.03	0.10	0.69	0.22	0.01	0.08
Ecology	0.65	0.09	0.01	0.25	0.63	0.06	0.01	0.29
Food/nutrition	0.49	0.36	0.03	0.12	0.44	0.35	0.00	0.21
Social sciences	0.53	0.04	0.01	0.41	0.57	0.03	0.02	0.39

Note: This table displays summary statistics by type of UIR and traditional scholarship. Proportions sum to 100 across columns. AE = academic engagement; AC = academic commercialization. We define traditional scholarship (TS) as those that do not engage in either AC or AE.

in commercialization activities led the way, with a 7 percentage point decline from 26 percent of respondents in 2005 to 19 percent in 2015. When we look at the four exclusive measures, this change concentrates on faculty moving from practicing both engagement and commercialization to engagement only. AE participation was essentially unchanged. The decline in AC between 2005 and 2015 contradicts the expected increase based on university-level commercialization promotion in previous decades. We conclude that the popular perception following university rhetoric on the expansion of UIR activities is not borne out by the behavior of LGU faculty in terms of engaging with industry in commercialization activities.

Across fields, participation varies between 60 and 95 percent of faculty engaging in any of the three types of UIR, as detailed in table 5.4. Although participation rates shown in the table are within each faculty characteristic, we also performed statistical tests to identify the differences in UIR participation across categories. There is statistically significant variation at a 95 percent level across gender, with men being on average 8 percentage points more likely to engage in any UIR than women. We find no statistically significant differences in participation by appointment type and/or level. In terms of field-of-study differences, the highest rates are in applied/production agricultural disciplines, while the lowest UIR participation rates are in the 60–70 percent range for the biological and social sciences. This outcome is also consistent with findings from the United Kingdom mentioned above, where more basic research is associated with relatively lower UIR activity.

While suggestive of different norms, the decline in commercialization

Table 5.5 **Persistence in faculty participation in UIR**

	2015				
	AE	AE/AC	AC	TS	Total
2005					
Academic engagement (AE)	99	27	2	18	146
AE and AC	16	30	2	4	52
Academic commercialization (AC)	1	2	1	2	6
Traditional scholarship (TS)	11	1	2	26	40
Total	127	60	7	50	244

This table reports results from the panel data linking individuals between 2005 and 2015 waves.

captured in the cross-section analyses might be a result of changes in the demographic composition of types of faculty. To control for potential demographic composition changes, we next investigate the individuals for which we have panel data among which demographic composition is constant. This smaller panel data set was gathered as part of the ongoing study to examine the persistence of individual participation in each of the categories. Table 5.5 provides a transition matrix between 2005 and 2015 of UIR participation rates across the categories.

We offer four observations based on the transition patterns in table 5.5. First, AE or mixed UIR categories show a higher rate of persistence over time than does commercialization only. There is a high exit rate out of AC reflected in the AE/AC and AC rows, where only a little over half of faculty that were doing commercialization in 2000–2005 stay engaged in AC activities in the 2010–15 time period. By contrast, about 85–90 percent of faculty who were engaged in AE or both activities in 2000–2005 remain engaged with AE activities in 2010–15. Second, the AC category is by far the least likely to gain faculty across the two time periods, reflecting the low likelihood of faculty activity in just commercialization. In fact, the decline in AC evident in the cross-sectional data also shows up as a lack of persistence and a lack of new faculty entrants into this activity. Third, a transition to AE/AC from any of the other categories is far more likely, suggesting the potential joint nature of AE with AC rather than the move to commercialization as an independent activity. Fourth, 25 percent of traditional scholars transitioned to AE activities over time, but at the same time, a larger number of scholars transitioned from UIR categories into the TS category. Thus the TS category increases from 16 percent to 20 percent of the sample, showing its robustness to the purported increase in UIR emphasis at LGUs.

Table 5.6 shows research funding for different UIR participation categories. It compares amounts of funding from different sources as well as the shares associated with each funding source.[7] Across all the UIR categories,

7. Note that "private industry" and "commodity organization" funding are used to define AE, and "patent royalties" is used to define AC. Therefore, by definition, these amounts are zero for some UIR categories.

Table 5.6 **Research lab financial sources across UIR types, 2005 and 2015**

		2005				2015			
		AE	AE/AC	AC	TS	AE	AE/AC	AC	TS
Research lab revs mean	$	155,491	213,848	197,625	107,411	293,202	403,127	346,602	271,649
Research lab revs median	$	75,000	150,000	101,500	60,000	100,000	200,000	250,000	60,000
Fed grants	$	89,900	112,497	157,860	77,634	180,415	238,995	274,314	223,796
	%	51.86	50.97	63.14	60.71	49.03	52.30	73.00	64.58
State grants	$	15,335	18,127	7,432	6,370	20,216	18,422	14,286	18,168
	%	9.16	6.25	7.32	8.06	8.14	5.18	2.86	5.22
Private industry	$	16,618	39,090	—	—	36,547	69,626	—	—
	%	11.96	17.63	—	—	12.67	17.15	—	—
Commodity orgs	$	7,385	10,623	—	—	19,348	24,674	—	—
	%	8.37	8.78	—	—	8.84	7.92	—	—
Foundations	$	6,016	9,339	6,398	7,087	12,326	17,228	35,232	15,828
	%	4.04	3.61	3.59	6.99	6.08	5.06	10.21	6.14
University funds	$	10,717	14,521	21,747	9,487	16,513	28,892	16,341	11,190
	%	10.90	7.74	17.50	15.19	11.05	9.44	13.21	13.49
Patent royalties	$	—	3,699	—	—	—	4,110	—	—
	%	—	1.06	—	—	—	1.42	—	—
Others	$	3,440	1,418	3,250	1,003	5,454	1,181	6,429	471
	%	2.08	1.02	5.31	2.51	1.18	0.55	0.71	1.56

federal funding remains the primary source of research funds, with industry and commodity organizations playing a substantial but subordinate role. At less than 2 percent overall, licensing revenues from AC activities are a trivial source, and they are one-tenth the value of the funds earned from private industry and one-third the value of funds from commodity organization sources. Interestingly, faculty who earned patent royalties are only found within the AC faculty who also engage in AE. It is also worth noting that for the median research lab revenue, those associated with faculty engaged in AC and AE/AC have the highest research funding levels across both years of data. Both AE-only and TS labs have lower levels of funding.

For each category of UIR, table 5.7 reports on scholarly outputs—namely, articles published in the last five years and being the main advisor for PhD and master's students. Consistent with many other previous studies in the literature, academic outcomes are robust to faculty participation in UIR activities. The most active faculty in UIR, the AE/AC group, have the highest article productivity (mean of 23 articles in 2010–15) and a similar number of PhD students trained (mean of 2.5 in 2010–15) to the AC group (2.6). These compare with about 14 articles over 2010–15 for AE and TS categories and 1.6 PhD students for those two categories. The high outputs of the AE/AC group are consistent with synergies between UIR and scholarly outputs that are found in econometric studies elsewhere (e.g., Foltz, Kim,

Table 5.7 Scholarly outputs across UIR types, 2005 and 2015

	2005							
	AE		AE/AC		AC		TS	
	Mean	Median	Mean	Median	Mean	Median	Mean	Median
Scholarly articles (5 yrs)	11.55	9	16.79	14	14.81	13	10.61	9
Master students (5 yrs)	3.08	2	2.61	2	1.94	1	2.95	1
PhD students (5 yrs)	1.22	1	1.75	1	1.88	2	1.72	1
	2015							
	AE		AE/AC		AC		TS	
	Mean	Median	Mean	Median	Mean	Median	Mean	Median
Scholarly articles (5 yrs)	14.41	12	23.45	19	21.36	16	14.37	11
Master students (5 yrs)	2.79	2	2.41	2	1.86	1	1.77	1
PhD students (5 yrs)	1.61	1	2.48	2	2.64	2	1.67	1

and Barham 2003). Table 5.7 is also noteworthy for providing continued evidence of rising productivity over time of US LGU faculty based on article counts (Prager, Foltz, and Barham 2014).[8]

We turn next to table 5.8, showing the values or stated preferences of US LGU faculty with respect to their motivations for "research problem choice." We first report for both 2005 and 2015 the average scores (1 low to 5 high) and compare them across UIR categories. In both years, "enjoy the research" and "scientific curiosity" scores average well above 4 for all categories of faculty. By contrast, the scores for "potential marketability" or "private firms commercialization interest" are lower for all the UIR categories relative to intrinsic motivations by at least a full point and oftentimes two or three points.

We next use factor analysis to recover underlying factors explaining the variance in the motivations for research choices data. Two factors explain most of the variance in the data, which we identify as intrinsic and extrinsic motivation factors (table 5.9). We constructed indexes (simple average) within the items identified as a factor. Some have consistent "high loadings" within each identified factor, such as scientific curiosity or potential contribution to scientific theory, which we interpret as intrinsic motivation. Meanwhile, likely interest by private firms in commercializing the discovery

8. We make no effort to control for quality or potential increases in coauthorship, either of which could lead to an adjustment in the raw measure provided here. The evidence from Foltz, Kim, and Barham (2003) suggests that quantity and quality (as measured by citations) are highly correlated, which means the bias from unmeasured quality could be small. We have no evidence on which way the bias from coauthorship patterns might go.

Table 5.8 **Research choice criteria across UIR types, 2005 and 2015**

	2005				2015			
Research choice criteria	AE	AE/AC	AC	TS	AE	AE/AC	AC	TS
Enjoy doing this kind of research	4.50	4.53	4.69	4.69	4.27	4.33	4.50	4.50
Potential contribution to scientific theory	3.49	3.78	4.38	4.13	3.37	3.73	4.50	3.83
Scientific curiosity	4.15	4.26	4.44	4.36	4.02	4.17	4.36	4.40
Probability of publication in professional journal	3.88	3.86	4.09	4.09	3.81	3.90	4.50	4.02
Potential marketability	2.42	3.35	2.69	1.64	1.77	3.06	2.21	1.36
Availability of private and corporate funds	2.88	3.35	2.03	1.71	2.84	3.48	2.00	1.89
Request made by clientele	3.28	3.32	2.06	2.09	3.09	2.97	1.64	1.76
Feedback from extension personnel	2.79	2.61	1.78	1.70	2.62	2.42	1.71	1.62
Potential to patent and license the findings	1.48	2.47	2.25	1.18	1.20	2.46	1.86	1.11
Interest by private firms in commercializing the discovery	1.79	2.76	2.09	1.25	1.44	2.66	1.71	1.15
Importance to society	4.31	4.27	4.28	4.21	4.06	4.27	4.29	3.95
Approval of colleagues	2.50	2.43	2.03	2.50	2.29	2.48	2.29	2.25
Availability of public funds	3.92	4.04	3.91	3.63	3.76	4.14	4.00	3.49
Availability of research facilities	3.46	3.83	3.25	3.11	3.28	3.89	4.14	2.86

Note: These questions are reported using a five-point Likert scale, with a score of 1 being "Not at all" and a score of 5 being "Extremely."

Table 5.9 **Factor loadings estimation, after rotation**

Item	Extrinsic	Intrinsic
Potential contribution to scientific theory		0.62
Probability of publication in professional journal		0.48
Enjoy doing this kind of research		0.51
Scientific curiosity		0.63
Request made by clientele	0.52	
Feedback from extension personnel	0.47	
Potential to patent and license the research findings	0.67	
Interest by private firms in commercializing the discovery	0.75	
Potential marketability	0.67	
Availability of private and corporate funds	0.52	
Availability of research facilities		
Approval of colleagues		
Availability of public, state, and federal funds		
Importance to society		

Note: Factors are calculated jointly for both waves. Comparing eigenvalues and their variances, we confirm the existence of two factors. Together, they explain 93 percent of the variance. We used principal factor with orthogonal quartimax rotation to estimate the factor loadings. Measures on intrinsic and extrinsic motivations are calculated as the average of the items within each identified factor.

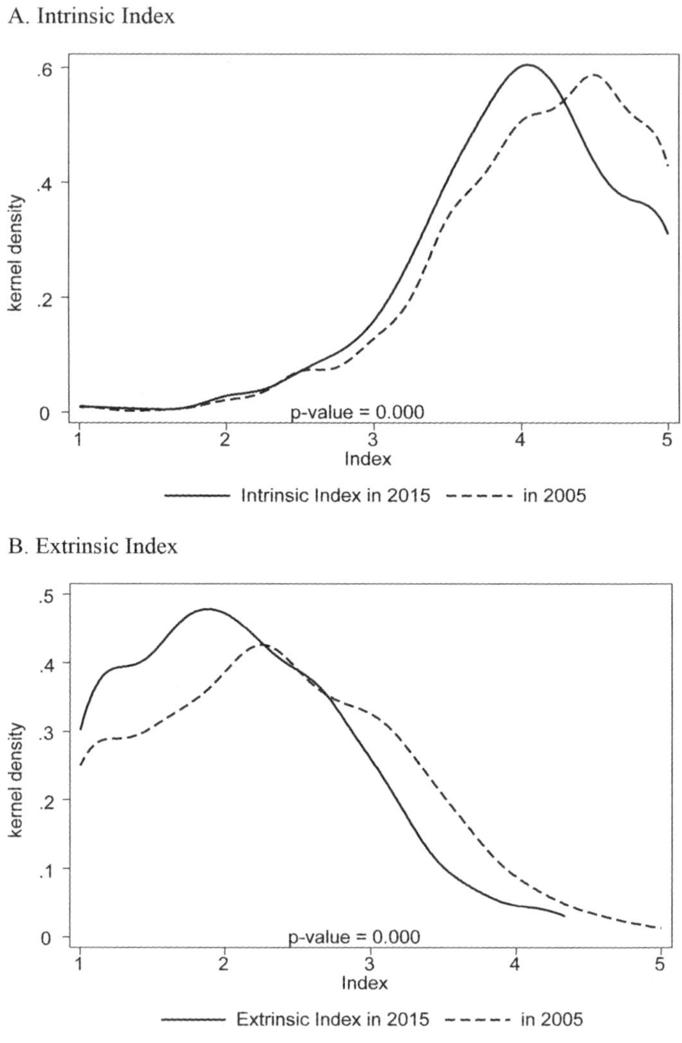

Fig. 5.1 Distribution of extrinsic and intrinsic incentives indexes, pooled cross-section data

Note: This figure displays the distribution of the calculated intrinsic and extrinsic indexes from the cross-section data for individual surveys in both 2005 and 2015.

and potential marketability of the final product "load high" in what we interpret as extrinsic motivation.

We show the distribution of the indexes by UIR category in figure 5.1. In 2005 and 2015, faculty report higher mean intrinsic than extrinsic motivations when it comes to research problem choice. The distribution of intrinsic motivations is skewed to the right, averaging 4 points. Meanwhile, extrinsic

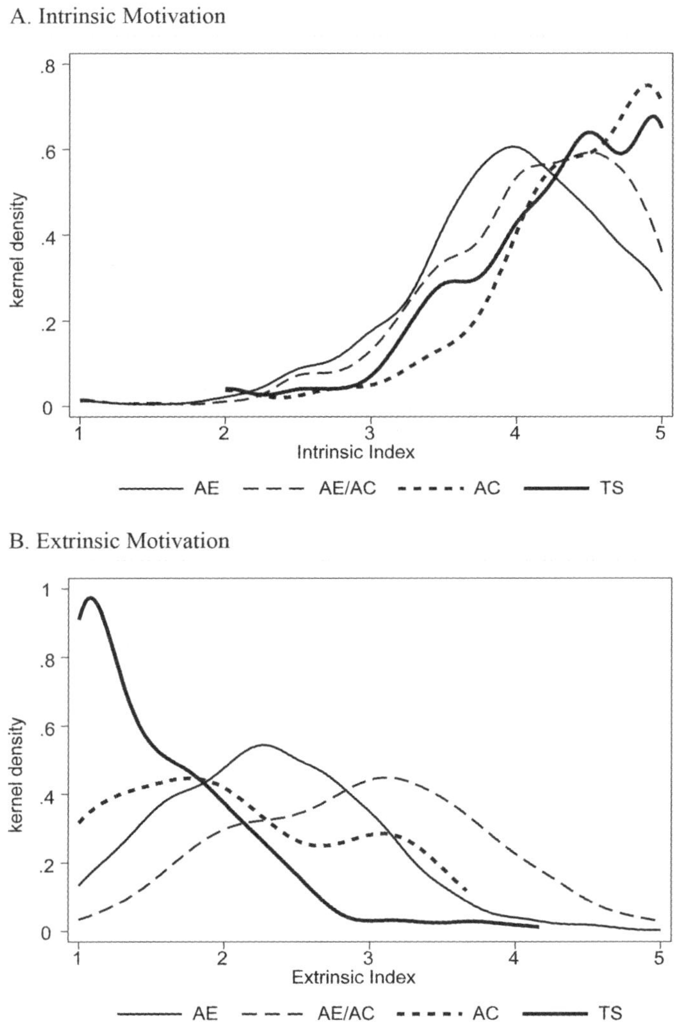

Fig. 5.2 Distribution of intrinsic and extrinsic motivation by UIR category, 2005 and 2015 pooled—index

Note: This figure displays the distribution of the calculated intrinsic and extrinsic indexes from the cross-section data for individual surveys in both 2005 and 2015 by UIR category.

motivation appears to be less important to faculty, averaging 2 points. Both measures decreased between 2005 and 2015, with the larger decrease in extrinsic motivation moving the distribution of that measure to be almost entirely distributed below "neutral." Meanwhile, the intrinsic index, while decreasing, remained strongly distributed in the very to extremely important zone.

In figure 5.2, we report the distributions of the motivation indexes by

UIR category. As would be expected by their actions, traditional scholars are skewed far to the right on intrinsic motivations and far to the left on commercial ones. Yet we see that both categories of commercialization (AC and AE/AC) also have high levels of scientific motivation, with only AE as distinctly below the others. AE/AC appears to show both the highest average levels of commercial motivation and the greatest diversity of motivations within the category, as exemplified by a flatter distribution. The AC-only group shows high levels of intrinsic motivation as well as a bimodal distribution of commercial motivation, with some at both high and low levels.[9]

5.4 Empirical Strategy and Results

Descriptive statistics show remarkable differences in academic outputs across UIR categories and in factors shaping research topic choice. In order to isolate these relationships, we use regression techniques to estimate correlations between UIR categories and various university outputs. The models, which should be interpreted as correlational rather than causal, use TS as our comparison category.

In the first set of estimates, we explore individual and institutional determinants of UIR engagement, including the relative role of faculty motivations, and field- and university-specific effects while controlling for faculty characteristics. We estimate equation (1) using a linear probability model with standard errors clustered at the university level to account for university-level heteroskedasticity. Our dependent variable UIR_{ifu} is a binary indicator variable for any UIR engagement, relative to TS. We adopt a flexible functional form to capture the potential correlation between motivation and UIR participation, $\Sigma_{k=1}^{k=4} Q_{k_i^m}$, with $m \in \{\text{Int}, \text{Ext}\}$. The regressors $Q_k F^m$ are indicators for each quartile k of each motivation F^m distribution. We omit the first quartile: $Q_1 F^m$. The vector X measures individual characteristics and includes gender, university appointment (professor, assistant professor, or full professor), and an indicator for whether the scientist was awarded a PhD from a land grant institution. The variables μ_f and ν_u are field and university fixed effects, respectively.

$$(1)\quad \text{UIR}_{ifu} = \alpha + \sum_{k=1}^{k=4} \beta_k^S Q_k F_i^{\text{Sci}} + \sum_{k=1}^{k=4} \beta_k^C Q_k F_i^{\text{Com}} + \gamma X_i + \mu_f + \nu_u + \varepsilon_{ifu}.$$

To demonstrate the correlations between UIR participation and our variables of interest, which are all categorical, we plot the effects in a series of figures. Figures 5.3 and 5.4 plot the set of estimated parameters for categorical variables β_k^m, μ_f, and ν_u, which are, respectively, motivation categories, university effects, and field effects.

9. Results for the exclusive AC category need to be interpreted with caution due to the small number of cases in our sample.

Figure 5.3 shows the parameter estimates for quartiles of the intrinsic and extrinsic motivation indexes on the probability that a faculty member engages in UIR activities. The figure shows estimated parameters for both unconditional (no other controls) and conditional (all controls in equation [1]) along with 95 percent confidence bands. The figure shows that as the intrinsic index increases, the probability of doing UIR activities marginally decreases, with intrinsic motivation playing a small role in differentiating UIR engagement. As for extrinsic motivation, there is a higher probability of UIR engagement as this indexes increases. These correlations corroborate the descriptive statistics that intrinsic motivation is high across the board, whereas extrinsic motivation plays an important role in differentiating UIR engagement. Overall, these determinants are robust to the inclusion of a variety of controls.

In figure 5.4, we show how the estimated parameters on university fixed effects, estimated from equation (1), vary across universities, with the University of Wisconsin-Madison (UW-Madison), which has the oldest technology transfer office among US universities, used as the baseline. There are relatively high university-specific effects, which indicate more UIR activity at that university compared to UW-Madison, at some of the large LGUs such as Illinois; the University of Massachusetts Amherst; the University of California, Davis; and Purdue. But we also see some smaller LGUs such as Alaska and Vermont in the top 15. There are some surprising effects with Cornell in the bottom tier along with a number of smaller LGUs that have fewer resources and newer traditions of UIR. Since we have controlled in these regressions for both individual- and university-observable characteristics, the best interpretation for these results is a measure of the UIR "culture" at these universities. Institutions such as Cornell may have stronger basic science cultures with less focus on UIR, while the large LGUs that are high on the list may have stronger outreach and extension cultures that promote more UIR.

The second half of the figure shows the estimated parameters on the field of specialty-level fixed effects, with plant sciences, which is the largest category, as the baseline. The other production agriculture sciences—namely, animal sciences, agricultural engineering, and food and nutrition studies—are not statistically distinguishable from plant sciences. This result, likely driven by academic engagement in production agriculture fields, is expected. Ecology and basic biological sciences, however, show lower levels of UIR engagement than do plant sciences, despite those fields possibly having higher potential in commercialization. This may be due to the stronger basic science orientation of these fields relative to applied production sciences. And as one would expect, the social sciences are at the lowest levels of all of the agriculture and life science college disciplines in terms of UIR activities.

In a second set of regression estimates, we isolate how each type of UIR activity correlates with academic productivity. The uniqueness of our data

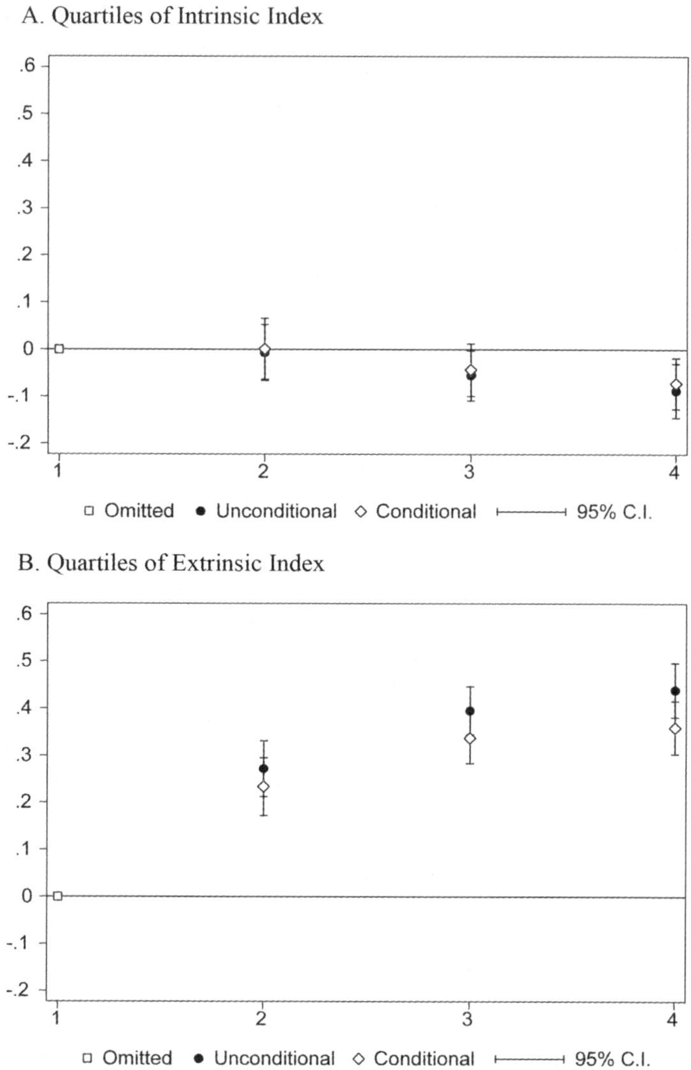

Fig. 5.3 Linear probability model: (Any) UIR engagement by quartile of attitudes

Note: Coefficients are for quartiles of motivation, with the first quartile as omitted variable. Dependent variable is an indicator for whether individual engages in any UIR type (1) as opposed to being a traditional scholar (0). Unconditional estimates include a survey year dummy. Controls for the conditional estimates include gender, position as professor, a dummy for whether PhD was in a land grant institution, field (plant science, ag/engineering, animal science, biology, ecology, food/nutrition, and sociology), and university fixed effects, which correspond to the 52 land grant universities. Standard errors are clustered at the university level.

A. University Estimated FE

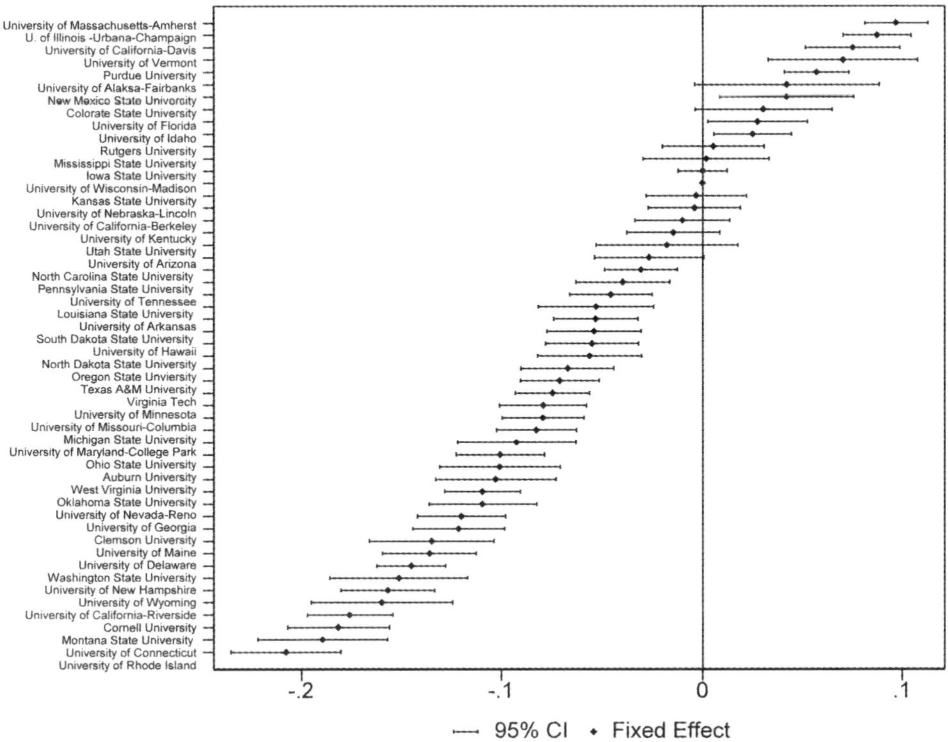

Fig. 5.4 Linear probability model of UIR engagement—field and university fixed effects

Note: For both panels, dependent variable is an indicator for whether individual engages in any UIR type (1) as opposed to being a traditional scholar (0). (a) Coefficients are for 52 university indicators, with University of Wisconsin-Madison as omitted variable. (b) Coefficients are for field indicators, with plant science as the omitted variable. We choose plant science as the omitted variables for being the most popular field in our sample. Additional controls include gender, position as professor, and a dummy for whether PhD was in a land grant institution. Standard errors are clustered at the university level.

set allows us to control for an often unobserved dimension of individual heterogeneity: faculty motivations, both intrinsic and extrinsic. This allows us to measure the direct effects of these characteristics beyond their effects that run through UIR engagement. We also control for individual, field, and institutional characteristics.

We estimate different versions of equation (2), in which Y_{ifu} varies in each regression covering the number of journal articles, PhD graduates, and total funding for scientist *i* in field *f* at university *u*. AC, AE/AC, and AE are our mutually exclusive measures of UIR, and TS is the omitted baseline category. The values F_i^I and F_i^E are the index scores for intrinsic and extrinsic

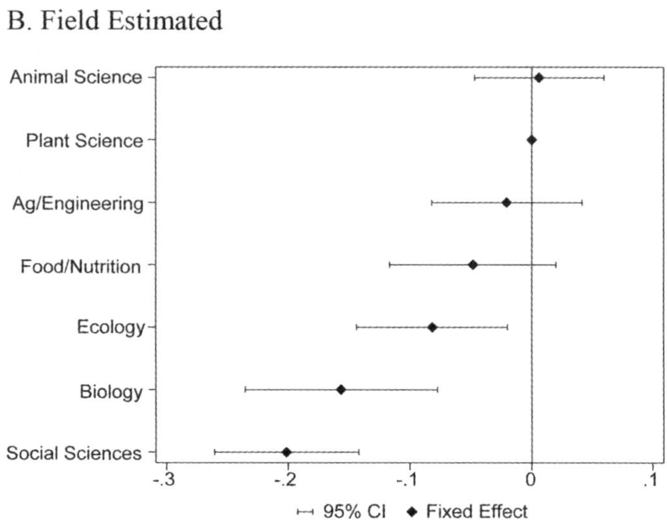

Fig. 5.4 (cont.)

motivation, respectively. The vector X measures individual characteristics and includes gender, university appointment (professor, assistant professor, or full professor), and an indicator for whether the scientist was awarded a PhD from a land grant institution. The variables μ_f and ν_u are field and university fixed effects, respectively. The standard errors are clustered at the university level to control for university-level heteroskedasticity.

$$(2)\ Y_{ifu} = \alpha + \beta_1 AC + \frac{\beta_2 AE}{AC} + \beta_3 AE + \psi_S F_i^S + \psi_C F_i^C + \gamma X_i + \mu_f + \nu_u + \varepsilon_{ifu}.$$

Table 5.10 shows the results of estimating equation (2) with journal articles and PhD students produced over the last five years as the dependent variable. The columns provide increasing levels of control variables: the first is the baseline, the second adds in our motivational measures, the third adds individual controls, and the fourth adds field and university fixed effects. One sees two dominant statistically significant and large effects: (1) compared to traditional scholars, AE/AC and AC-only faculty produce more journal articles and more PhD students, and (2) levels of intrinsic motivations are directly correlated with both journal articles and PhD students, whereas extrinsic (commercial) motivations do not play a direct role besides those embedded in how they determine UIR participation. Overall, the picture that emerges from table 5.10 is that UIR faculty, especially those with commercial ties, are more productive than traditional scholars. In addition, those in the AE-only category appear to produce scholarship and students at about the level of traditional scholars.

Table 5.11 shows the results of estimating equation (2) using the amount

Table 5.10 **OLS estimates—journal articles publications and PhD graduates under supervision, 2005 and 2015 pooled**

	Journal articles				PhD graduates			
AE only	0.556	1.635*	1.588*	0.543	−0.321**	−0.070	−0.072	0.070
	(0.82)	(0.84)	(0.75)	(0.15)	(0.15)	(0.14)	(0.14)	(0.14)
AE/AC	7.155***	7.625***	6.902***	4.881***	0.316*	0.611***	0.449**	0.545***
	(1.13)	(1.05)	(1.03)	(1.11)	(0.18)	(0.20)	(0.19)	(0.17)
AC only	5.061***	4.403**	4.440**	3.857**	0.445	0.466	0.510	0.693*
	(1.8)	(1.81)	(1.80)	(1.63)	(0.37)	(0.38)	(0.36)	(0.36)
Intrinsic motivation		3.921***	4.241***	3.868***		0.405***	0.470***	0.447***
		(0.47)	(0.49)	(0.50)		(0.08)	(0.08)	(0.08)
Extrinsic motivation		0.201	0.337	0.282		−0.160**	−0.105	−0.041
		(0.42)	(0.41)	(0.42)		(0.07)	(0.07)	(0.06)
Survey year; individual controls; field/university FE	x	x	x x	x x x	x	x	x x	x x x
Observations	1,479	1,479	1,479	1,479	1,479	1,479	1,479	1,479
R-squared	0.070	0.110	0.153	0.223	0.024	0.049	0.174	0.271
p-value F-test (UIR = 0)	0.00	0.00	0.00	0.00	0.187	0.011	0.055	0.007

Note: Coefficients on UIR categories are relative to traditional scholars (omitted). AE = academic engagement; AC = academic commercialization. Dependent variables are total of articles published in the last five years and number of PhD graduates under supervision in the last five years. Individual controls include gender, position as professor, and a dummy for whether PhD was in a land grant institution. Field includes plant science, ag/engineering, animal science, biology, ecology, food/nutrition, and sociology. University fixed effects correspond to the 52 land grant universities. Standard errors are clustered at the university level. * $p < 0.10$, ** $p < 0.05$, *** $p < 0.01$.

Table 5.11 **OLS estimates, total and public funding, 2005 and 2015 pooled**

	Total funding (IHS)				Public funding (IHS)			
AE only	1.174***	1.369***	1.383***	1.337***	1.039***	1.555***	1.558***	1.573***
	(0.27)	(0.28)	(0.28)	(0.27)	(0.34)	(0.37)	(0.37)	(0.35)
AE/AC	2.080***	2.111***	2.109***	1.964***	1.822***	2.368***	2.316***	2.288***
	(0.27)	(0.30)	(0.31)	(0.30)	(0.36)	(0.40)	(0.40)	(0.39)
AC only	1.724***	1.554***	1.536***	1.362***	1.320*	1.306**	1.309**	1.288*
	(0.41)	(0.38)	(0.38)	(0.39)	(0.68)	(0.63)	(0.64)	(0.71)
Intrinsic motivation		0.858***	0.846***	0.715***		0.996***	1.013***	0.916***
		(0.11)	(0.12)	(0.11)		(0.18)	(0.18)	(0.19)
Extrinsic motivation		0.096	0.081	0.197*		−0.261*	−0.252*	−0.151
		(0.09)	(0.09)	(0.10)		(0.14)	(0.14)	(0.16)
Survey year; individual controls; field/university FE	x	x	x x	x x x	x	x	x x	x x x
Observations	1,540	1,540	1,540	1,540	1,540	1,540	1,540	1,540
R-squared	0.059	0.102	0.107	0.161	0.023	0.056	0.059	0.100
p-value F-test (UIR = 0)	0.00	0.00	0.00	0.00	0.00	0.00	0.00	0.00

Note: Coefficients on UIR categories are relative to traditional scholars (omitted). AE = academic engagement; AC = academic commercialization. Dependent variables are current annual budget and its subcategory of total public funding. Total public funding categories are USDA, NSF, NIH, other federal agencies, and state agencies. Individual controls include gender, position as professor, and a dummy for whether PhD was in a land grant institution. Field includes plant science, ag/engineering, animal science, biology, ecology, food/nutrition, and sociology. University fixed effects correspond to the 52 land grant universities. Standard errors are clustered at the university level. * $p < 0.10$, ** $p < 0.05$, *** $p < 0.01$.

of total funding and public funding as two dependent variables. It is worth noting that our categories of UIR are partially created with funding data, so we should expect a positive relationship with total funding, though not with public funding. Here we see very strong and statistically significant correlations of any UIR activity with both total funding and federal funding. While the former is somewhat expected, the latter suggests that rather than being a distraction from TS directions, faculty engagement in UIR activities is synergistic in terms of bringing in federal funding, which is generally associated with TS. Again, we see strong correlations of intrinsic motivations with both total and federal funding, suggesting a direct effect, whereas extrinsic motivation does not have such a direct effect.

5.5 Discussion

Our empirical results present a number of new findings for the study of UIR activities at US universities on several important fronts. First, faculty participation rates in UIR activities are quite high; generally, around 70–80 percent of US LGU agricultural and life scientists engage in AE, AC, or both. Second, faculty participation in UIR is predominantly in the area of AE, the more traditional type of research collaboration involving sponsored research, industry collaboration (including farmers and their commodity organizations), and other types of research exchanges (presentations and shared problem identification). In fact, only about 2–3 percent of faculty in either the 2005 or 2015 survey participated in just AC activities. Third, as a source of research funding for ALS faculty at US LGUs, AE industry revenues completely dominate AC license revenues, but the largest individual faculty funding levels come from those who do both AE and AC. Overall, patent license revenues provide about 1 percent of lab revenues as compared to approximately 25 percent share of lab revenues coming from industry and commodity group funds. This funding outcome appears to be in "steady state" now 35 years after the passage of the Bayh-Dole Act and more than 25 years after the takeoff of US public university patenting activity, as the ratio of AE to AC funding was the same in 2015 as it was in 2005.

This study also finds descriptive evidence that UIR activities are highly correlated and likely synergistic with traditional academic scholarship activities. This outcome is consistent with previous studies that find the more productive researchers are also often the ones most highly "in demand" or active in UIR activities. While this study does not undertake the type of longitudinal dynamic statistical analysis as do Sengupta and Ray (2017), who find positive feedbacks between AE and research outcomes at the university level, prima facie evidence presented in our work at the individual faculty level is consistent with that outcome. In particular, our finding that the AE/AC faculty persist across time periods and that this group has more research revenues and higher publication and student counts demonstrates this individual positive feedback loop.

In examining factors shaping the participation of US LGU faculty with UIR activities, we find that institutional factors, specifically "fields" or "disciplines," are a significant conditioning factor, with more applied science fields such as plant and animal sciences having higher UIR rates than more basic ones such as biological and ecological ones. Additional analyses show that most of the differences in UIR activity by field are driven by variations in AE rather than AC. This finding is consistent with both the lower overall participation in AC and the fact that most of the faculty active in AC are also active in AE. The reverse is not true. Most faculty engaged in AE are not active in AC. In this regard, it appears that AC may be somewhat opportunistic and may depend on the types of inventions or discoveries being made by scientists. Put simply, ongoing collaboration with industry or sponsored research arrangements may, from time to time, give rise to the pursuit of invention disclosures and patents, and so entry and exit into AC activities appear to occur regularly, as shown in the transition matrix in table 5.5.

The most substantive individual factors shaping the intensity of participation in UIR appear to be faculty "attitudes" with respect to research problem choice. While we do not attempt here to identify a causal relationship between attitudes and UIR activity involvement, ALS faculty at US LGUs report that their research problem choices are strongly driven by intrinsic factors, such as curiosity or the potential to contribute to scientific theory relative to intrinsic motives. This is true across all the UIR categories used here, though what distinguishes AE, AC, and AE/AC from TS is a somewhat stronger level of extrinsic motive. This basic preference for science has been a consistent outcome across decades of surveys of US LGU faculty and is consistent also with the continued importance of federal competitive grants as a primary source of research funding.

Finally, university fixed effect measures in our UIR regressions reveal statistically significant differences in university "cultures" with respect to UIR. These differences appear to relate to the timing of initial commercialization activity and potentially to other historical and locational factors that could be important for how they shape faculty behavior over time. This is an area of ongoing interest and potentially productive future inquiry.

5.6 Conclusion

This chapter has examined UIR activities of ALS faculty at the premier US LGUs, using survey data gathered from large, random, and longitudinal samples in 2005 and 2015. The analysis of this unique set of data fills an empirical gap identified in the literature by carefully exploring the relative importance of AE and AC. Because US LGUs are the "ground zero" of US public research university UIR activities, the empirical context is of broader significance to the United States and beyond. We have found descriptive and correlational evidence that traditional academic scholarship has not systematically been distorted or constrained in the ways that some originally feared and that UIR—while important to faculty, universities, and society—is not a fundamental threat to the advancement of science.

At US LGUs, the long-standing tradition of AE, involving sponsored research and direct collaboration with scientists and managers in industry and agriculture, dominates the new AC relationships in prevalence and importance for faculty research funding. Moreover, these two types of UIR appear to be complements, with AC being an occasional outgrowth of AE in some fields and for some faculty, which likely depends on the continuity of AE relationships to emerge. Seen in this way, the UIR activities of agricultural and life scientists at LGUs are more of a natural outgrowth of the land grant system's traditional model of working with industry to foster improved outcomes in their own states and the nation. Fears of UIR subverting the LGU mission appear to be misplaced. Rather, we find that UIR complements the traditional work of top scholars in ALS fields in part by helping them access more funding and connections with industries in their field.

The data described and analyzed in this article represent a valuable resource for investigators who seek to understand the workings of ALS research in the United States. Future exploration with these data will seek to pursue causal identification of UIR participation and intensity outcomes using historical information as instruments as well as more exploitation of the panel data. Expanding the focus on university-level factors seems worth special attention in this effort. Adding to the current data set with measures of journal quality would also be a useful contribution. In addition, given the significant growth in the proportion of women and foreign faculty in the US LGUs over time, there are open important questions as to whether this has changed the dynamics of UIR participation. Using the data described in this chapter to analyze how the ALS research establishment has or has not diversified over the last quarter century, and the effects thereof on research output and topics, is an important avenue for understanding the conduct of science.

Appendix

Sample Selection and Imputation of Missing Values

Within the subsample of individuals who completed the survey, there was a large number of missing responses. We assumed a set of hypotheses in order to impute the missing values. (1) Research attitudes: Likert scale ranging

Table 5.A1 **Summary of response rates for the random and panel samples and subsample selection criteria**

	2005	2015
Full sample	2,330	2,972
Valid responses	1,590	977
Response rate (%)	68	32
Random sample	1,328	711
Panel sample	262	266
Panel matched	**244**	
Random sample 2005–15	2,039	
drop field=other,missing	19	
drop not professorship, missing	83	
drop cross-missing	397	
Final subsample (random sample)	**1,540**	

from 1 to 5. We assigned a neutral value, 3, if the individual answered the block at least partially. When all responses are missing, variables remain coded as missing. (2) UIR-related measures: individual responses with missing values are replaced with 0 when the person answered part of the block. When all individual responses for the questions within the block are missing, we do not input values. (3) Extension and outreach: missing responses are coded as 0 if the block was partially answered. If all questions within block were not answered, variables remain coded as missing. (4) PhD students: missing responses are coded as 0 if the block was partially answered. If all questions within the block were not answered, variables remain coded as missing. For each block, we created variables indicating the total number of imputed values, and results are robust to adding these variables in the regression as a control. These results are available upon request.

References

Agrawal, A. 2001. "University-to-Industry Knowledge Transfer: Literature Review and Unanswered Questions." *International Journal of Management Reviews* 3:285–302.

Agrawal, A., and R. M. Henderson. 2002. "Putting Patents in Context: Exploring Knowledge Transfer from MIT." *Management Science* 48:44–60.

American Academy of Arts and Sciences. 2016. *Public Research Universities: Understanding the Financial Model.* Cambridge, MA: American Academy of Arts and Sciences.

Azoulay, P., W. Ding, and T. Stuart. 2007. "The Determinants of Faculty Patenting Behavior: Demographics or Opportunities?" *Journal of Economic Behavior and Organization* 63:599–612.

Barham, B. L., J. Foltz, M. I. R. Agnes, and J. van Rijn. 2017. "Modern Agricultural Science in Transition: A Survey of US Land-Grant Agricultural and Life Scientists." Staff Paper 584, University of Wisconsin-Madison.
Barham, B., J. Foltz, and K. Kim. 2002. "Trends in University Agbiotech Patent Production." *Review of Agricultural Economics* 24:294–308.
Barham, B., J. Foltz, and D. Prager. 2014. "Making Time for Science." *Research Policy* 43:21–31.
D'Este, P., and M. Perkmann. 2011. "Why Do Academics Engage with Industry? The Entrepreneurial University and Individual Motivations." *Journal of Technology Transfer* 36:316–39.
Dillman, D. A. 2011. *Mail and Internet Surveys: The Tailored Design Method—2007 Update with New Internet, Visual, and Mixed-Mode Guide*. Hoboken, NJ: John Wiley & Sons.
Djokovic, D., and V. Souitaris. 2008. "Spinouts from Academic Institutions: A Literature Review with Suggestions for Further Research." *Journal of Technology Transfer* 33:225–47.
Ehrenberg, R. 2012. "American Higher Education in Transition." *Journal of Economic Perspectives* 26:193–216.
Fitzgerald, H. E., K. Bruns, S. Sonka, A. Furco, and L. Swanson. 2012. "The Centrality of Engagement in Higher Education." *Journal of Higher Education Outreach and Engagement* 16:7–28.
Foltz, J., K. Kim, and B. Barham. 2003. "A Dynamic Analysis of University Agricultural Biotechnology Patents." *American Journal of Agricultural Economics* 85:187–97.
Geuna, A., and A. Muscio. 2009. "The Governance of University Knowledge Transfer: A Critical Review of the Literature." *Minerva* 47:93–114.
Goldberger, J., J. Foltz, B. Barham, and T. Goeschl. 2005. *Summary Report: Modern Agricultural Science in Transition: A Survey of US Land-Grant Agricultural and Life Scientists*. PATS Research Report 14, University of Wisconsin-Madison.
Grimaldi, R., M. Kenney, D. Siegel, and M. Wright. 2011. "30 Years after Bayh-Dole: Reassessing Academic Entrepreneurship." *Research Policy* 40:1045–57.
Henderson, R., A. Jaffe, and M. Trajtenberg. 1998. "Universities as a Source of Commercial Technology: A Detailed Analysis of University Patenting, 1965–1988." *Review of Economic Statistics* 80:119–27.
Hoag, D. L. 2005. "WAEA Presidential Address: Economic Principles for Saving the Cooperative Extension Service." *Journal of Agricultural and Resource Economics* 30 (3): 397–410.
Just, R. E., and W. E. Huffman. 2009. "The Economics of Universities in a New Age of Funding Options." *Research Policy* 38:1102–16.
Mitchell, M., V. Palacios, and M. Leachman. 2015. "States Are Still Funding Higher Education below Pre-recession Levels." *Journal of Collective Bargaining in the Academy* 0 (10). https://www.cbpp.org/research/states-are-still-funding-higher-education-below-pre-recession-levels.
Perkmann, M., R. Fini, J. M. Ross, A. Salter, C. Silvestri, and V. Tartari. 2015. "Accounting for Universities' Impact: Using Augmented Data to Measure Academic Engagement and Commercialization by Academic Scientists." *Research Evaluation* 24 (4): 380–91.
Perkmann, M., Z. King, and S. Pavelin. 2011. "Engaging Excellence? Effects of Faculty Quality on University Engagement with Industry." *Research Policy* 40:539–52.
Perkmann, M., V. Tartari, M. McKelvey, E. Autio, A. Brostrom, P. D'Este, R. Fini et al. 2013. "Academic Engagement and Commercialisation: A Review of the Literature on University-Industry Relations." *Research Policy* 42:423–42.

Phan, P., and D. Siegel. 2006. "The Effectiveness of University Technology Transfer." *Foundations and Trends in Entrepreneurship* 2:77–144.
Prager, D. L., J. D. Foltz, and B. L. Barham. 2014. "Making Time for Agricultural and Life Science Research: Technical Change and Productivity Gains." *American Journal of Agricultural Economics* 97:743–61.
Sampat, B. 2006. "Patenting and US Academic Research in the 20th Century: The World before and after Bayh-Dole." *Research Policy* 35: 772–89.
Sengupta, A., and A. S. Ray. 2017. "University Research and Knowledge Transfer: A Dynamic View of Ambidexterity in British Universities." *Research Policy* 46: 881–97.
Tartari, V., M. Perkmann, and A. Salter. 2014. "In Good Company: The Influence of Peers on Industry Engagement by Academic Scientists." *Research Policy* 43: 1189–203.
Tartari, V., and A. Salter. 2015. "The Engagement Gap: Exploring Gender Differences in University-Industry Collaboration Activities." *Research Policy* 44: 1176–91.
Thursby, J., and M. Thursby. 2011. "Has the Bayh-Dole Act Compromised Basic Research?" *Research Policy* 40:1077–83.

Comment Nicola Bianchi

This chapter explores the characteristics of university-industry relations (UIR) among agricultural and life science (ALS) faculty at US land grant universities (LGUs). This research question is interesting because there is a common belief that US universities are relying more and more on UIR to replace dwindling funding from the state and federal government. Despite the plausible growing importance of UIR activities, little is known about their features, their links with professors' academic output, and their consequences for academic research.

This chapter contributes to our understanding of UIR in several ways. It uses extensive survey data collected in 2005 and 2015 to explore UIR at LGUs. The sample is large, covering 946 professors in 2005 and 626 professors in 2015. Among these faculty members, 234 are surveyed in both years, allowing the analysis to have a panel component. Moreover, the survey asks detailed questions about UIR, permitting the authors to distinguish between different forms of UIR. Specifically, the chapter is able to differentiate between academic engagement (AE) and academic commercialization (AC). AE describes any form of faculty participation in shared research.

Nicola Bianchi is an assistant professor of strategy at Northwestern University and a faculty research fellow of the National Bureau of Economic Research.

For acknowledgments, sources of research support, and disclosure of the author's material financial relationships, if any, please see https://www.nber.org/books-and-chapters/economics-research-and-innovation-agriculture/comment-academic-engagement-commercialization-and-scholarship-empirical-evidence-agricultural-and.

It involves, for example, research support from industry, participation in industry presentations, and research collaborations with industry experts. AC describes any form of faculty participation in private intellectual property creation. For example, it involves the creation of patents, products, or start-ups with industrial partners.

This chapter adopts different methodologies to describe the characteristics of UIR. It compares differences in the average characteristics of faculty members participating in different types of UIR. Moreover, it creates a UIR index via principal factor analysis. Finally, it uses multivariate regressions to concurrently control for multiple drivers of UIR.

Before moving to the discussion of the main results, it should be noted that this chapter accomplishes a lot. It produces an impressively extensive analysis on UIR. It answers at least six different and important research questions on the topic. Moreover, the chapter uses newly available survey data. These data are great for many reasons. First, the sample size is big enough to obtain precise estimates. Second, this survey is designed to ask very detailed questions on UIR. As a result, the survey allows the authors to study correlations that could not be addressed by previous research on this topic. Given the importance of the data, I think that future work should put more emphasis on them. It should be clearer that the data represent a significant contribution to the literature. Moreover, future work should include more information on the survey itself, especially on the 2005 wave. In addition, any upcoming research should discuss the representativeness of the sample. It would be very informative to have a table in which the faculty members in the sample are compared to all other ALS professors at US LGUs. Ideally, the average characteristics of these two groups will be statistically similar. In addition, I think that a natural way to expand the data would be to incorporate administrative data on patent production and on publications instead of relying exclusively on self-reported output measures. The addition of these data sets would be valuable for at least two reasons. First, any self-reported variable raises questions about its reliability. Administrative or third-party data would assuage these concerns. Second, these types of administrative data would include objective measures of the quality of the academic production, which are missing from the current analysis.

Moving to the discussion on the findings, the first main result shows that between 80 and 90 percent of surveyed faculty members participate in UIR. Moreover, there is little variance across gender and academic rank. There is, however, significant variance across academic fields. The UIR rate varies from 94 percent in animal science to only 68 percent in social sciences. In future work, I would emphasize a bit more this last finding (differences across fields). Furthermore, it would be very beneficial to draw a tighter connection between the results and the hypotheses outlined in section 5.3. For example, I think it would be better to state that the differences across fields are consistent with hypothesis 7. In addition, it would be beneficial to remind the reader that the high average participation rate is consistent with hypothesis 1.

The second main result is that AE is much more prevalent than AC. In 2005, 55 percent of faculty members participated in only AE activities, while only 3 percent participated in only AC activities. Among the remaining faculty members, 23 percent participated in both, while 19 percent did not participate in UIR at all. This result is interesting because it might speak to the nature of AC and AE activities. Is it possible that AC is, in most cases, the second or advanced phase of UIR after an initial period limited to AE activities? This could be true because most collaborations might start solely as research support. Sometimes, research activities (AE) are successful and open the path for further collaboration on commercialization (AC). This pattern could explain why AE is more widespread. Moreover, it would explain why almost nobody engages exclusively in AC activities.

The third main result shows that UIR participation fell by 3 percentage points between 2005 and 2015. Over this period, the largest shifts were away from AE/AC combined and from AC only. Instead, the participation rate in AE only increased by 7 percentage points. These results are very interesting because they partially contradict the hypothesis that states that UIR activities are on the rise. The truth is that only a subgroup of UIR, AE activities, has been increasing over the last 15 years. I think that future research could estimate the same changes in UIR using a multinomial logit model. This type of model would account for the fact that the outcomes (participation in UIR) are mutually exclusive and would produce more robust estimates.

The fourth main result shows that funding for UIR came predominantly from federal and state grants. Moreover, the importance of government grants increased over time. Possibly contrary to popular belief, patent royalties are not a substantial stream of revenues. On average, they contribute around 1 percent of revenues for faculty members participating in AC activities. These results are very interesting because they partially contradict the idea that dwindling government grants are one of the driving forces of UIR. Although it is true that state grants have been decreasing between 2005 and 2015, federal grants increased over time. It is also true that revenues from private industry have been increasing. A hypothesis is that UIR activities provide significant funding for US LGU research activities. Based on historical trends in AE and the recent push for the expansion of AC activities, along with declines in state funding levels, UIR is expected to play a significant and perhaps growing role in funding faculty research activities. These results, however, paint a more complex picture. Therefore, I think that future research should have a more nuanced discussion on how these results relate to the original hypothesis. Moreover, although the main finding is quite clear (strong reliance on federal funding), I have some doubts about other sources of funding. "Private industry" is a very general label for a funding source. The same comment applies to "Foundations." Would it be possible to dig deeper into these sources and unpack their overall contribution into smaller subgroups? I think that doing so would help with interpretation. It would be especially important for "Private industry," considering that it is a funding source that is becoming more important over time.

The fifth main result shows a positive correlation between UIR and scholarly output. Specifically, faculty members engaged in UIR have an average academic production that is higher, compared with faculty members not engaging in UIR. Moreover, the scholarly production is the highest for professors who engage in both AE and AC. This result corroborates the hypothesis that "UIR activities are broadly synergistic with other US LGU outputs such as producing articles and training graduate students." I think that this result is fascinating because it is the first step toward addressing what might motivate professors to engage in UIR. Future research should dig deeper into these findings. First, I think that this analysis requires a multivariate regression. In fact, a regression could allow the authors to estimate the correlation between UIR and scholarly output while also controlling for other extraneous factors. Moreover, it would allow the authors to assess whether differences across UIR are statistically significant. Second, as mentioned above, adding external data on academic activity (e.g., the Web of Science Data by Thomson Reuters) would make it possible to measure the quality of the scholarly output instead of focusing only on quantity. Beyond these technical issues, I have some comments on the interpretation of these interesting results. The chapter states that these findings are consistent with the idea that there are "synergies" between UIR and academic activities. What is the true meaning of *synergies* in this context? Is *synergy* just a synonym of *correlation*, meaning that in the data, the two activities are more likely to happen together? Or does *synergy* imply an actual mechanism?

The sixth main result shows that scientific motivations to engage in UIR are more important than commercial ones. These findings are fascinating but come with some caveats. First, the motivations for UIR are self-reported. This is not necessarily a limitation of the data, because there is no other way to collect this information. However, self-reported data could be skewed toward finding that scientific motivations are more important than any other factor. There is likely a strong negative stigma attached to faculty members who identify commercial motivations as their primary driver. I think that future research should discuss this fact. Second, these results are only partially corroborating the authors' claims. Specifically, a hypothesis is that the pursuit of scientific discoveries is the primary motivation shaping US LGU faculty participation in UIR activities. I think that the results in table 5.8 do not necessarily prove this point. In other words, they do not prove that scientific motivations are the primary motivation behind UIR. These findings prove that scientific motivations are more widespread and common than commercial motivations. Third, I would emphasize how commercial motivations are more common among UIR faculty. Specifically, the share of faculty members who have commercial motives is often more than 1 percentage point higher among faculty members engaging in UIR. Fourth, as I explained in the previous paragraph, this analysis should be performed using multivariate regressions.

The seventh main result shows how a UIR index, built using principal component analysis, is correlated with scientific and commercial motivations. I agree with the idea of building a UIR index, because engaging in UIR activities is likely a continuous choice, not a dichotomous one. Specifically, the chapter creates two indices: one index for AE and one for AC. The main issue is that a change in these two indices implies different comparisons. As shown in figure 5.1, an increase in the AE index compares faculty members not engaging in UIR to faculty members engaging almost exclusively in AE. Instead, an increase in the AC index compares faculty members engaging in AE to faculty members engaging in both AE and AC. These discrepancies make the interpretation of these last results a little complex.

To conclude, this chapter represents one of the most thorough explorations of UIR to date. The analysis of newly available survey data dispels some common misconceptions about UIR. Specifically, the chapter suggests that there are stark differences between AE and AC activities. As a consequence, every discussion on this topic should take into account that the various types of UIR have very different characteristics. I suggest that future research addresses the following four main issues. First, it would be beneficial to assess the representativeness of the survey data with respect to all ALS professors at US LGUs. Second, follow-up work should discuss in greater detail the complex relationship between trends in government grants (federal vs. state) and participation in UIR activities. Third, the analysis of the effects of UIR activities on scholarly output should be based on multivariate regressions in order to control for confounding factors. Moreover, future work should add third-party administrative data on academic publications and patent production (such as the Web of Science database) in order not to rely on self-reported outputs exclusively. Fourth, the UIR indexes should be redesigned to make them more comparable.

6

Venture Capital and the Transformation of Private R&D for Agriculture

Gregory D. Graff, Felipe de Figueiredo Silva, and David Zilberman

6.1 Introduction

Innovation in the agricultural and food system has been fundamental in enabling it to feed the world. Developments in mechanical, chemical, and biological technologies have contributed to productivity gains that have more than doubled outputs of agricultural production over the last 50 years while scarcely changing the aggregate quantity of inputs (Alston et al. 2010). Innovations in harvesting, processing, and other postharvest steps have also increased the capacity and efficiency of the food system, helping improve food security and the nutritional quality of diets for a growing global population (FAO 2018).

Innovation in modern agriculture increasingly occurs as a result of formal research and development (R&D) activities, conducted in both the public sector and the private sector. Historically, agricultural R&D has been highly managed. In the mid-19th century, it was led by governments supporting agricultural research stations and research at agricultural colleges and universities. By the mid-20th century, an international agricultural research system, supported and overseen by philanthropic foundations and international organizations, became a major source of new innovations. During

Gregory D. Graff is a professor in the Department of Agricultural and Resource Economics at Colorado State University.

Felipe de Figueiredo Silva is an assistant professor of agribusiness in the Department of Agricultural Sciences at Clemson University.

David Zilberman is the Robinson Chair Professor in the Department of Agricultural and Resource Economics at the University of California, Berkeley.

For acknowledgments, sources of research support, and disclosure of the authors' material financial relationships, if any, please see https://www.nber.org/books-and-chapters/economics-research-and-innovation-agriculture/venture-capital-and-transformation-private-rd-agriculture.

this same time frame, large corporate agribusiness and food firms increased their R&D with the objective of maximizing the profitability of their core production and marketing activities.

While government investments in agricultural R&D have been declining in real terms in high-income countries over the last several decades, industry investments in agricultural R&D have increased steadily (Fuglie et al. 2012; Pardey et al. 2016). Globally, annual industry expenditures on agricultural R&D in 2009 were in the range of $10 billion (Fuglie et al. 2011) to $16 billion (Pardey et al. 2015). The most recent available global estimate of industry's agricultural R&D was $15.6 billion in 2014 (Fuglie 2016). However, all such estimates primarily consider publicly listed companies that are subject to public disclosure requirements by securities regulators. Such estimates have largely ignored small- and medium-size enterprises (SMEs) because historically, they have contributed very little to industry R&D.

In recent years, however, there has been a surge in the founding and financing of start-up companies seeking to develop and apply new technologies in agriculture and the food system. These companies are privately held and have raised significant amounts of equity-based investment from venture capital (VC) funds and related private sources such as seed, angel, and other private equity investors. According to industry reports, in recent years, up to several billion dollars annually have been invested into such agricultural technology (or "agtech") start-up companies (AgFunder 2015, 2019; CBInsights 2017; Dutia 2014; Finistere Ventures 2019; KPMG 2018). While the phenomenon of start-up companies or new technology-based firms (NTBFs) introducing new technologies to agriculture is itself not new, recent rates appear to be unprecedented in terms of both the number of start-ups and the amount being invested in them.

Yet these various accounts of R&D investment in agriculture draw on a range of different private data sources and industry subsector definitions and thus vary in terms of how prevalent they find agtech start-ups to be and how much VC they find is being invested in the industry. To some extent, this variation is due to the inherent challenges of industry classifications. Established categories tend to reflect the incumbent structure of industry—such as seeds, agricultural chemicals, or agricultural machinery. However, many of the recent agtech start-ups span conventional industry categories. For example, a start-up may have its primary industry classification in software, yet that software may be highly specialized for data management and decision support of on-farm crop production. One perennial question is the extent to which downstream food manufacturing, wholesale, and retail categories should be included and how, especially since the business models of some of today's leading start-ups explicitly seek to shorten or span the entire "farm-to-table" value chain. Variations in accounting of VC investments are also due to the fact that, historically, private investments in agricultural R&D have been quite low in developing countries (Pardey and Beintema

2001; Pardey et al. 2006). Yet recently, robust start-up activity and private investment is being reported in middle- and lower-income countries, especially in the larger emerging economies like India (AgFunder 2018b), China (AgFunder 2018a; Gooch and Gale 2018), and Brazil (Mondin and Tomé 2018; Dias, Jardim, and Sakuda 2019). Data sources and procedures for systematic compilation of small-scale private business activity in such countries are nascent at best. It is not clear why this surge in venture investment in agricultural technology has occurred in middle- and lower-income countries now or what factors account for this apparent upturn, but it appears to be an important part of this global phenomenon and has remained largely unrecorded.

We present evidence in this chapter that until 2006, the amount invested globally in agtech start-ups remained relatively negligible, typically less than $200 million per year, then grew steadily from 2007 to 2009 and then exploded in 2010, exceeding $3 billion annually in recent years. One industry source claims that venture investments in agricultural technology may have been as high as $7 billion in 2018 (AgFunder 2019). At that rate, VC and associated private investors could be allocating up to half as much toward agricultural R&D as are the corporate members of the industry.

These start-ups and their R&D activities can be expected to impact existing agricultural technologies and industry structure. These start-ups are tapping new sources of financing to support R&D for agriculture. Compared to established R&D organizations, in both the public and private sectors, venture-backed start-ups are subject to different incentives and constraints and are connected to different professional networks. This allows them, collectively, to pursue a larger and more diverse range of R&D projects. Some of the R&D conducted by start-ups may be complementary to R&D by established organizations. Some are even spun off from established R&D organizations to build on discoveries made within those organizations in order to transfer or translate those findings into market applications. Other start-ups are contributing new research tools or platform technologies—such as novel sensor systems, artificial intelligence algorithms, or genome editing technologies—that could improve the research productivity of all agricultural R&D organizations, public and private. Yet other start-ups may be directly competing with established public sector or corporate R&D agendas, seeking to "disrupt" current technologies or ways of doing business.

The VC-backed start-up is effectively a mechanism to contain the financial risks of prospecting in the process of R&D, reducing or managing the technical and market uncertainties of innovation. While many start-ups fail in the attempt, some do prevail in bringing their innovation to market. An increase in the rate of successful start-ups may help counter recent trends of increased market concentration in agribusiness, in which fewer larger firms have been accounting for ever greater shares of private sector R&D (Fuglie 2016). VC-backed start-ups bring Schumpeter's "gale of creative destruction," supplanting some current technologies and companies. Without innovation, market concentration can lead to exploitative monopolies, but with innovation, new competition can erode monopoly power (Zilberman, Lu, and Reardon 2019).

This chapter investigates the increase in the number of new agricultural technology start-ups globally. What are the dynamics of entry and growth of new firms financed by VC? Where is it occurring? To what extent are they concentrated in high-income countries? And what are the main market categories or technologies they are pursuing?

This chapter also explores a range of economic factors and circumstances that might help explain this growth of VC investments in agriculture. A better understanding of the factors causing this investment will help us anticipate whether it is merely a transient phenomenon or whether it constitutes a more enduring shift in the composition and dynamics of agricultural R&D. Other industries, such as software, internet services, and pharmaceuticals, have both enjoyed exponential growth and endured downturns in venture investment, most famously with the bursting of the tech bubble circa 1999–2000. Yet today, those sectors continue to exhibit an innovation ecosystem that is routinely refreshed by new start-ups funded by VC in an ongoing virtuous cycle. The fundamental question is the extent to which the R&D and innovation system of agriculture is being transformed by this influx of equity-based private investments in R&D-intensive start-up companies and whether it will come to operate more like these other high-tech industries in the long run.

To investigate these changes, we compile a global data set of 4,552 companies in agriculture, founded from 1977 to 2017, with 11,998 associated financial transactions, including investments into and exits from these start-ups. The lack of reporting requirements for privately held firms generally makes it difficult to systematically track start-ups and their financing (Cumming and Johan 2017). To overcome this challenge, we draw primarily from the commercial data vendor PitchBook (by Morningstar) and augment its data with additional company and financial transaction records from competing commercial data sources, VentureSource (by Dow Jones) and Crunchbase (founded by TechCrunch). The financial transactions reported include a range of VC, seed, and angel investments; some other private equity deals; and debt financing. They also include transactions by which early investors and founders exit their investments in these start-ups, such as initial public offerings (IPOs), mergers and acquisitions (M&As), and other types of buyouts. While the transactions data do indicate some bankruptcies and closures of the start-up firms, the reporting of these is incomplete, and so we are left to impute a rate of firm closures based on clues in the data. Together, these data allow us to explore the start-up life cycles and exit outcomes over time and across the full range of different technologies being developed (e.g., biotech vs. software), across the full range of subsectors of agriculture (e.g., inputs vs. outputs, or crops vs. livestock), and across the globe.

Our data summaries show exponential growth in the number of start-ups from about 2009 to the present. The largest share of start-ups is in the United States (33 percent), followed by Europe (23 percent), with the remainder (44 percent) elsewhere in the world. Significant numbers are in emerging and developing economies, such as India (5 percent), China (4 percent), and Brazil (2 percent). In terms of technologies being developed, about one-third of the new start-ups involve computer, information technology (IT), and data-related technologies; another third involve biotechnology, breeding, genetics, or animal health; and the final third encompass a wide range of other technologies, applications, and business models, including marketing and sales, financial and business services, and even on-farm production.

This chapter proceeds as follows. We turn next to a quick overview of the economic literature on agricultural R&D and on VC. We then introduce a new data set on agricultural technology start-ups. The full sample of start-ups is used to track overall trends, such as founding rates, the geography of start-ups globally, and start-ups by technology or industry categories. A narrower subset of start-ups that also have data on their investments is used to analyze growth in investments and factors associated with that growth, both at the firm level and at the industry level. The results suggest that recent surges in commodity prices—together with higher amounts of VC being invested overall and signals from successful exits in agriculture—may have led to the rise in VC investments in agriculture. We conclude that VC investments into start-up companies represent an important new source of R&D expenditures with the potential to transform many aspects of private R&D for agriculture.

6.2 Literature Review

6.2.1 Financing of R&D in Agriculture

There is a robust agricultural economics literature on the institutional and financing aspects of agricultural R&D (Alston et al. 2010; Huffman and Evenson 2006; Pardey, Alston, and Ruttan 2010; Sunding and Zilberman 2001). Relative to other industries, agriculture has long had a high ratio of public sector to private sector R&D. Pardey and Beintema (2001) tracked spending globally over several decades and estimate that in 1995, total global agricultural R&D was $33.2 billion, of which 65 percent ($21.7 billion) was by public sector sources (defined as research conducted by or funded by governments, academics, or nonprofit organizations), while 35 percent ($11.5 billion) was by the private sector (defined as profit-motivated R&D by privately or publicly held companies and state organizations). Five years later, in 2000, global total spending on agricultural R&D was only slightly higher, at $33.7 billion, and the sectoral shares had adjusted slightly, with the share conducted by the public sector down to approximately 60 percent and the share conducted by the private sector up to around 40 percent (Pardey et al. 2006).

Several key trends have been observed in the composition of agricultural R&D globally. The share of global agricultural R&D conducted in middle- and low-income countries is about 45 percent versus 55 percent conducted in high-income countries, which is a much higher share than overall R&D conducted in low- and middle-income countries, which is 22 percent versus 78 percent in high-income countries (Pardey et al. 2015). However, of the agricultural R&D conducted in low- and middle-income countries, very little of it is in the private sector. Historically, private sector R&D in developing countries is very low: in 1995, of the agricultural R&D conducted in developing countries, only 5.5 percent was by the private sector (Pardey and Beintema 2001).

Over the last two decades, agricultural R&D has grown steadily but unevenly. In the United States and other high-income countries, public sector spending is growing only very slowly in nominal terms and has declined in real terms (Pardey et al. 2016). At the same time, public sector spending has surged in middle-income countries, particularly in China (Hu et al. 2011). Private sector R&D has grown steadily in both high-income and middle-income countries. Private expenditures on agricultural R&D in 2009 were on the order of $10 billion (Fuglie et al. 2011) to $16 billion (Pardey et al. 2015), with differences in the estimates depending largely on which industry subsectors of the agricultural and food system are included or how data for unobserved spending by SMEs is estimated (Fuglie 2016). The most recent available global estimate of private sector agricultural R&D was $15.6 billion in 2014 (Fuglie 2016). At the same time, private sector agricultural R&D has become increasingly concentrated in the hands of fewer, larger companies (Fuglie et al. 2011).

Such accounts, however, have been based primarily on R&D spending by publicly listed companies. It has not been feasible nor, frankly, relevant to be concerned about R&D spending by SMEs, including VC-backed companies. While biotechnology start-ups were observed to have contributed significantly to the rise of genetic engineering in agriculture in the 1980s and 1990s (Fuglie et al. 2011; Fuglie 2016; Graff, Rausser, and Small 2003), levels of R&D spending and other financial data on such privately held companies are relatively inaccessible, as they are not subject to the same reporting requirements as publicly traded firms. Moreover, the relative amounts of R&D spending contributed by SMEs have historically been negligible (Fuglie 2016).

6.2.2 Venture Capital Investments

Dixit and Pindyck (1994) developed the standard methodology used to assess investment decisions, taking uncertainty and irreversibility into account. They argue that while the net present value approach is meaningful when considering whether to make an investment at a given moment in time, in most realistic situations, investors also have to decide about the timing of their investment and therefore have to take into account the randomness of key variables such as costs. The timing of an investment is thus triggered when the key random variable exceeds a certain threshold, also known as a hurdle rate. A good example of this approach in agriculture is the uncertainty around investing in new irrigation technologies due to agricultural prices and weather uncertainty (Carey and Zilberman 2002). Farmers only adopt new irrigation technologies when prices exceed a certain threshold.

The same basic logic can be applied to VC investments in agricultural technology start-ups. Even though VC investments have been feasible for decades, it was only after 2010 that they increased significantly (see figure 6.4). Several factors may have affected the hurdle rate, such as an increase in the ratio of agricultural prices to nonagricultural commodity prices, the occurrence of large exit events in highly visible start-ups, the emergence of new technological opportunities based on advances in enabling technologies (such as genome sequencing, genome editing, or data capacity of sensors and networks), and changes in (agricultural) labor markets in both high-income and middle-income countries.

In general, it has been shown that the dynamics of VC markets are driven by several measurable factors, including expected investment returns, the overall health of the economy, industry characteristics, and company financial performance variables (Gompers and Lerner 2004). VC funds that invest in agriculture are no different. Fundamentally, they are seeking returns on investment. Investors compare performance across industries, aspiring to identify high expected returns. Large positive swings in agricultural commodity prices would be expected to shift the supply of VC investments toward start-ups in this industry. Changes in commodity prices such as those observed especially between 2007 and 2014 might have played a role in the increase of the supply of VC investments in agriculture, even though Deloof and Vanacker (2018) observe that Belgian start-ups founded during the 2007 crisis had a greater chance of facing bankruptcy. In examining economic determinants of VC funding, Groh and Wallmeroth (2016) and Jeng and Wells (2000) investigate both developed countries and emerging markets. Groh and Wallmeroth (2016) show that the global share of VC investments in emerging markets increased from 2.4 percent in 2000 to 20.8 percent in 2013, indicating that the salient factors for VC investors are increasingly found in emerging markets.

Gompers and Lerner (2004) point out the greater number of rounds and larger amounts of VC investments go into high-tech industries, such as computers and biotechnology, compared to other more traditional industries. Puri and Zarutskie (2012) compare VC- and non-VC-financed firms and find that the key firm characteristic that attracts VC investment is its potential for scale. Even though agriculture, broadly speaking, may be considered a traditional industry with low margins, most VC investments in the sector are targeting the application of high technologies, such as geospatial technologies, digital sensors, robotics, biotechnology, automated vertical farming, alternative protein products, artificial intelligence–driven decision-making tools, and big data for supply chain management (AgFunder 2015; Graff, Berklund, and Rennels 2014; Rausser, Gordon, and Davis 2018). Regulations can influence investments in agricultural technologies as well. For example, regulations imposed by different countries or regions (such as the European Union) on gene editing might lead to big changes in biotech investments, with potential market uptake depending on whether other countries will follow the European or the American regulatory standards for this technology (Rausser, Gordon, and Davis 2018).

6.2.3 Venture Capital Exits

There is a growing literature examining exit outcomes as a key factor in the functioning of venture capital markets. Large exit events, including IPOs and M&As of start-ups may foment further investments. There is evidence on the positive effect of the size of IPO exits (Jeng and Wells 2000) and M&A exits (Félix, Pires, and Gulamhussen 2013; Groh and Wallmeroth 2016) on subsequent VC investments in earlier-stage start-ups. In agriculture, the acquisitions of the Climate Corporation by Monsanto in 2013 for $930 million and of Blue River Technology by John Deere in 2017 for $305 million were widely publicized and may have stimulated subsequent investments by VCs in other agricultural technology start-ups.

The literature investigating start-up exits identifies key factors that affect both new company starts and existing companies' survival, such as real interest rates, other macroeconomic variables, company sizes, and industry-specific variables (Holmes, Hunt, and Stone 2010; Giovannetti, Ricchiuti, and Velucchi 2011). Audretsch (1994, 1995) also shows that such variables can in turn determine the exit outcome—finding, for example, that start-up size is related to chance of exit, while industry growth rate is not. Puri and Zarutskie's (2012) comparison of VC-backed and non-VC-backed companies finds evidence that companies with VC investors have a higher likelihood of resulting in an M&A or IPO exit and a lower likelihood of a failed exit, controlling for industry-specific characteristics and year fixed effects. Gompers and Lerner (2004) have extensive discussions on the likelihood of start-ups going public via IPO, and they show that generally better industry conditions, such as those captured in an industry equity index (e.g., biotechnology index), are positively associated with that industry's number of IPOs.

Previous and contemporaneous exit outcomes, even in emerging and developing economies, are found to be directly associated with VC investments. While Groh and Wallmeroth (2016) find evidence that M&As impact

VC funding positively, Jeng and Wells (2000) find that IPOs play a greater role in determining VC investments in the later stages of the start-up life cycle. Investments into technologies that may be related to the agricultural industry are also location specific (Kolympiris and Kalaitzandonakes 2013; Pe'er and Keil 2013; Kolympiris, Kalaitzandonakes, and Miller 2015; Kolympiris, Hoenen, and Kalaitzandonakes 2017). This, combined with observations that overall agricultural R&D activities have shifted toward emerging markets, makes it reasonable to expect that the share of VC investments in agriculture has shifted toward emerging markets as well.

6.2.4 Venture Capital and Innovation

Following results by Kortum and Lerner (2000), which suggest that VC dollars may be three times as productive as corporate R&D dollars in generating patents, a number of studies have examined the relationship between VC and innovation. The hypothesis that VC-backed firms are more innovative is consistent with more general observations that VC investors select firms that are more likely to succeed and to do so at scale (Baum and Silverman 2004; Engel and Keilbach 2007; Puri and Zarutskie 2012), but there is also evidence that VC investors encourage companies in which they invest to enhance their knowledge absorption and R&D capacity (Da Rin and Penas 2017). There is evidence that start-ups receiving VC investments file more patent applications both in the short run and more permanently, and moreover, those patent applications are more likely to be granted, an indication of higher-quality innovation (Arqué-Castells 2012).

Within the population of VC-backed start-ups, there may be higher payoffs for those that are more innovative. Nadeau (2011) finds that VC-backed start-ups that exit via the more profitable IPO route are more likely to be engaged in patenting than those that exit via M&A, at least in key sectors such as biotechnology, IT, and internet services. Gaulé (2015) similarly finds that VC-backed start-ups with higher-quality patents are more likely to be successful, exiting via an IPO or a highly valued M&A.

One question that arises is the extent to which private equity or VC investment into start-up companies can be compared to or even proxy for R&D expenditures. Kortum and Lerner (2000) and Metrick (2007) distinguish between R&D financed by corporations and R&D financed by VC. However, publicly traded corporations report R&D expenditures according to established requirements, while small privately held firms do not. Kortum and Lerner (2000) assume that the bulk of venture financing goes to support innovative activities while acknowledging that some VC investments may be made in low-technology start-ups or may be spent on other activities such as marketing. Whether these exceptions are greater in agriculture than in industries that have traditionally received VC investment is an important but ultimately empirical question.

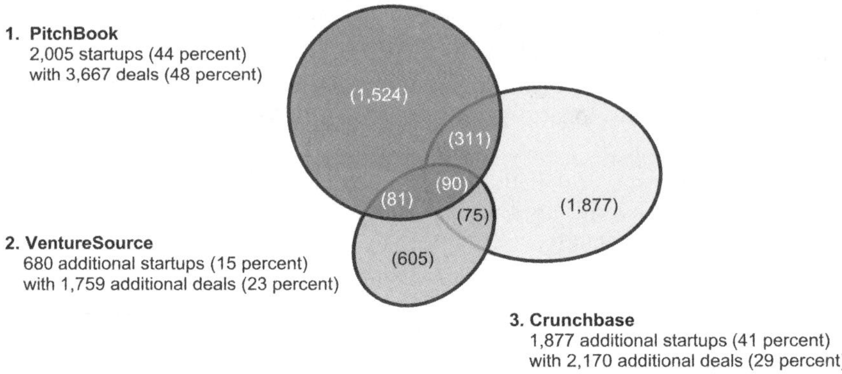

Fig. 6.1 Venn diagram of data accessed on new start-ups in agriculture and their financial deals by data source

Note: Our primary data source is PitchBook, augmented by additional company and financial deal records from VentureSource and then Crunchbase; data for just 12 percent of total start-ups were found to be available from more than one source.

6.3 Data on Venture Capital Investments in Agriculture

The data for this first look at VC investment in agriculture is drawn from several commercial data sources and consists of information on 4,552 start-up companies and 7,596 financing deals. We follow the approach of Hall and Woodward (2010) to compile a data set drawn from a variety of sources in order to overcome the limitations of data reporting and potential biases of any one source. Of the sources we draw on, the industry standard is generally regarded to be PitchBook, a financial database focused on VC and private equity investing, owned by Morningstar. Data from PitchBook are then augmented with data from two other sources: VentureSource and Crunchbase. From each source, two types of data are available, linked in a one-to-many relationship: data on companies and data on financing deals of those companies.

A comparison of company data listings across these three sources was undertaken with an expectation that overlap among data sources would allow for the cross-validation of firms and their deals. However, as figure 6.1 illustrates, we find minimal overlap of company listings across these sources. Our initial sampling, drawn from PitchBook, included 2,005 companies founded in the 40 years from 1977 to 2017, along with 3,667 financial deals for those companies, as designated by PitchBook's "AgTech Vertical" industry category.[1] From VentureSource, by Dow Jones, we drew an additional

1. PitchBook (n.d.) defines an industry vertical or vertical market as "a more specific industry classification" that "identifies companies that offer niche products or fit into multiple industries" or that represents "new fields with promising companies that attract investors." PitchBook describes the agtech vertical as consisting of "companies that provide services, engage

680 companies with 1,759 associated deals—identified by VentureSource's "Agriculture and Forestry" industry category[2]—that were not found in the PitchBook data. From Crunchbase—identified by their "Agriculture and Farming" industry group[3]—we drew an additional 1,885 companies beyond those listed in either PitchBook or VentureSource and 2,170 associated deals for those companies. Just 557 (12 percent) of the total companies were found in more than one data source, and only 90 (2 percent) of the total companies identified were listed in all three sources.[4] This pattern of data availability suggests that any analysis based on one primary data source (e.g., AgFunder 2015, 2019; CBInsights 2017; Finistere Ventures 2019; KPMG 2018) provides only a limited and largely separate sampling of overall venture investments in the industry.

Of the total 4,552 unique companies and 7,596 unique deals, we take about half of the data records on companies and deals in our collection from PitchBook and the other half from VentureSource and Crunchbase (figure 6.1). Of these sources, PitchBook had the most complete data overall, VentureSource was more complete in reporting older companies and deals, and Crunchbase was helpful in identifying a wider range of start-up companies internationally, but unfortunately, it was not able to provide as much coverage of deal information for those firms. Overall, deals data are associated with only 1,584 (35 percent) of the companies in the combined data set. Of the subset of companies with deal data, 1,366 (29 percent of the total) report at least one deal in which the amount is disclosed (others report deals that occurred but do not disclose amounts), and 1,092 (24 percent) report an identified exit deal.

Given these discrepancies in the availability of deal information, the subsequent analysis is undertaken in two parts. First, we summarize major industry trends using the full data set of 4,552 companies. Second, we summarize investments and exits for the 1,348 start-ups with accompanying deals data that disclose amounts, and we analyze those factors that may be associated with the recent growth in those investments. Arguably, given the skewed nature of valuations and investments in VC markets generally,

in scientific research, or develop technology which has the express purpose of enhancing the sustainability of agriculture. This includes wireless sensors to monitor soil, air and animal health; hydroponic and aquaponic systems; remote-controlled irrigation systems; aerial photo technology to analyze field conditions; biotech platforms for crop yields; data-analysis software to augment planting, herd, poultry and livestock management; automation software to manage farm task workflows; and accounting software to track and manage facility and task expenses."

2. VentureSource's "Agriculture and Forestry" industry category is a subset within its larger category of "Industrial Goods and Materials."

3. Crunchbase's "Agriculture and Farming" industry group includes companies in agriculture, agtech, animal feed, aquaculture, equestrian, farming, forestry, horticulture, hydroponics, and livestock.

4. For those 12 percent listed in more than one data source, for each company we use only data from one source, depending on availability, in the following order of preference: (1) PitchBook, (2) VentureSource, (3) Crunchbase. See the numbers of companies and deals in figure 6.1.

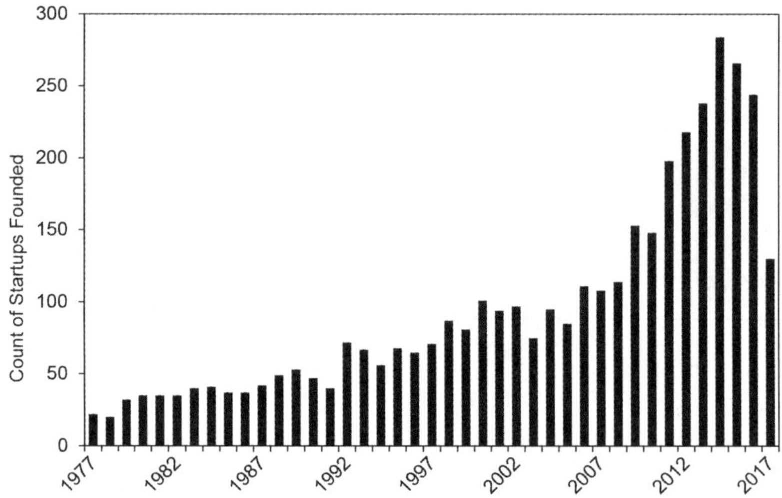

Fig. 6.2 VC-funded start-ups in agriculture by founding year, 1977–2017
N = 3,891 companies for which founding year is reported or proxied by first deal date

together with a propensity to report information on larger, more significant investments and exits (Hall and Woodward 2010), it stands to reason that the 29 percent of companies with disclosed deals represents a greater share of the overall financing activity in the industry. It is important to emphasize that despite efforts to be inclusive, this data set is still necessarily an under-representation of overall activity in the industry. Yet it provides a broad, representative sampling of private investment activity across agriculture globally.

6.3.1 Full Combined Data Set of Start-Up Companies in Agriculture: Global Summary Statistics

For many of the 4,552 companies in the combined sample, the founding date is available. For those companies with the founding date missing but for which deal information is available, we use the date of the first deal as a proxy for the founding year. Figure 6.2 plots the number of start-ups by founding year.

Qualitatively, there appear to have been three phases of growth in agricultural start-ups. First, from 1977 to 1991, we see steady, slow growth, with between 20 and 50 start-ups globally each year (however, this time period precedes the full data coverage for some of the countries and/or data sources on which this data set draws). In the second phase, from 1992 to 2008, the growth rate appears to have increased, yet it also appears to be more volatile—characteristic of the wider tech sector during this period—

with a downturn for several years following the bursting of the tech bubble in 2000. Third, starting in 2009—a year after grain prices reached a peak associated with the expansion of the US biofuel mandate (Wright 2009)—growth in start-ups experienced a sharp increase that, arguably, continues until the end of the time frame of this analysis despite the right-hand truncation seen here.[5]

The overall sample of 4,552 companies also includes data on physical address, which allows us to analyze the geographical distribution of entrepreneurship in agriculture globally. We find 1,483 start-ups in the United States, which accounts for about 33 percent of the global sample (figure 6.3). Within the United States, by far the most are in the state of California (348), with other leading states including Colorado, New York, Texas, Massachusetts, and Illinois. Of the US start-ups, 320 were located across the 11 midwestern states that encompass the highly fertile Corn Belt region. The European Union has 1,063 start-ups, accounting for about 23 percent of the global total, led by the United Kingdom (with 261), France (173), and Spain (102). Canada is home to 228 start-ups (5 percent of the global total). Among the emerging markets, India stands out with 210 start-ups, followed by China (172). South American countries have 144 start-ups in the sample, led by Brazil (88). Agricultural start-ups are also found in Africa, with the most in South Africa (41), followed by Kenya (31) and Nigeria (27). The pattern of VC-funded start-ups follows the growth pattern of VC in developing countries identified by Groh and Wallmeroth (2016). This global distribution of start-ups appears to track somewhat more closely with the pattern of public sector agricultural R&D, with a significant share in emerging and developing economies (Pardey et al. 2016), compared to the pattern of private sector agricultural R&D, which is more heavily concentrated in high-income countries (Fuglie 2016).

The categorization of start-up companies by industry—or of innovations by technology field—has long been a fraught exercise. Of the three data sources, each provides several data fields describing the company, its market activities, and the technologies it is developing. However, the descriptions of companies are very heterogeneous, even within the same data field from the same source. Even standardized industry category variables, which are more consistently reported within each data source, are not readily comparable across the three data sources. We therefore construct a common categorization for the start-ups in the combined sample, drawing on the full range of these descriptive data fields across all three sources based on

5. The apparent decline in start-ups after 2014 is, arguably, due to truncation in the data. New companies generally get reported to these data sources upon their first formal equity-based investment, which can occur up to several years following the founding of the company. Industry reports such as AgFunder (2019) show steady continued growth in start-up activity through 2018.

A. Global Density Mapping, by City and/or Postal Code

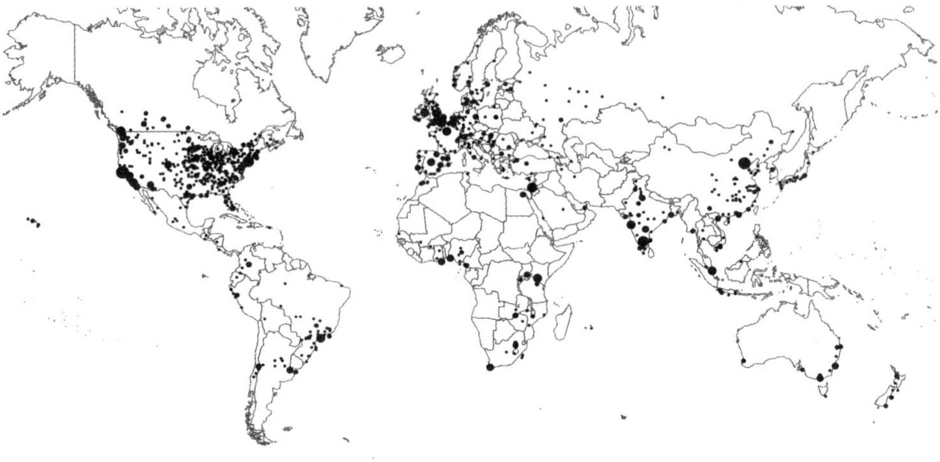

B. Global Share, by Country and Region

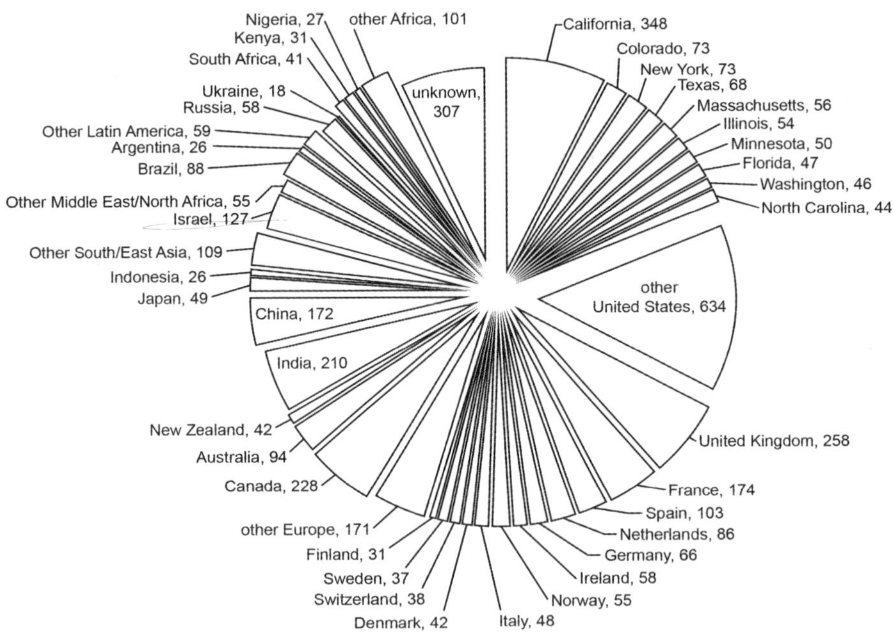

Fig. 6.3 The global scope of new VC-funded start-ups in agriculture, 1977–2018

Note: (a) Global density mapping by city and/or postal code and (b) global share by country and region

Table 6.1 **Industry and technology categories characterizing venture-funded start-ups in agriculture**

Category*	Number of start-ups with activities described by each category
Agricultural input technologies and services	2,482
Software, data, and information technologies	942
Electronic devices, sensors, and systems (electronic hardware)	430
Genetics, breeding, biotech, and health	918
Chemicals	230
Machinery and equipment (mechanical hardware)	302
Agricultural input distributions and sales	678
Agricultural production and farming	467
Agricultural output marketing, processing, and manufacturing	730
Consumer products and services	105
Business and financial services	539
Online services and content	471
Unspecified	1,165

N = 4,552 firms, of which 1,226 (27 percent) are identified with two or more categories

industry observations (see Graff, Berklund, and Rennels 2014; Dutia 2014; and AgFunder 2019), as detailed in the appendix.

It is important to note that the categories we develop are not exclusive. By their very nature, start-ups often span more than one industry or technology. Of the 4,552 start-ups in the data set, 1,226 (27 percent) span more than one category in table 6.1. Of these, 1,048 start-ups are classified in two categories, 161 in three, and 15 in four. For example, we have a start-up that is developing sensors with specialized data management tools to conduct high-throughput phenotyping to decipher crop genetics. Such a firm would be labeled with three of these categories: (1) electronic devices, sensors, and systems; (2) software, data, and information technologies; and (3) genetics, breeding, biotech, and health. While such an approach does result in multi-counting of firms by categories, it is not an uncommon practice.[6]

Table 6.1 displays the number of start-ups described by each of the categories we developed. Just over half of the start-ups in the data set are involved in some form of agricultural input technology or service, which in turn encompasses a number of different technology-based subcategories. Of these, the two largest are software and data (which describes 942 start-ups) and biotech, genetics, or health (which describes 918 start-ups). Companies identified by one or both of these categories attracted more than 60 percent of the venture investments made in the industry in 2016.

6. For example, under the International Patent Classification (IPC) system, multiple patent classifications can be assigned to a single patent.

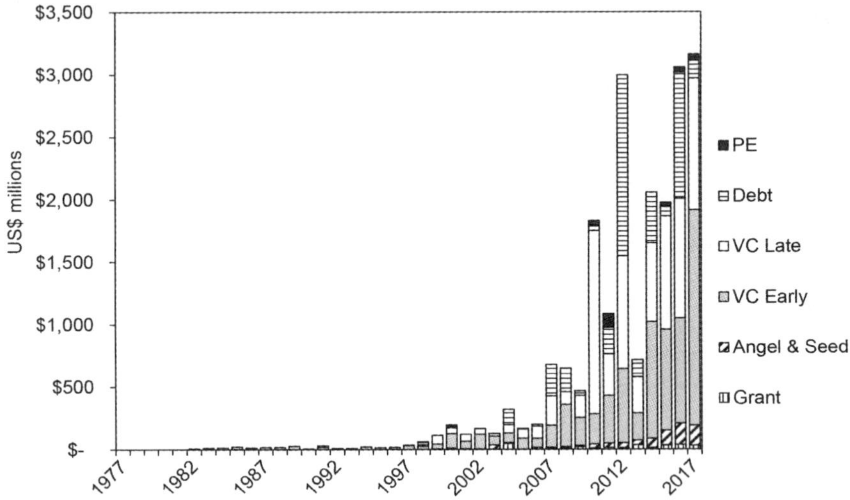

Fig. 6.4 Investments into VC-funded start-ups in agriculture by type of investment, 1977–2017

Note: PE = private equity; VC = venture capital

6.3.2 Subsample of Companies with Reported Deals: Investments and Exits

Of the 4,552 companies in the overall data set, only 1,584 (35 percent) are associated with the 7,596 reported deals (which, again, include investments, successful exits, and reported closures or bankruptcies). However, of these reported deals, many do not disclose the amount of the deal. To analyze VC investment trends, we narrow our information down to a subsample of just those 1,367 start-ups (29 percent of the overall sample) that report at least one venture-type money-in investment with a disclosed amount. In other words, all additional money-in investments received by that same start-up are considered in this analysis, including grants, angel, and seed stage investments; early-stage VC, late-stage VC; debt; and any other private equity investments. Companies that did not report deal amounts and companies with only private equity investments or debt financing were dropped from the sample for this part of the analysis. Figure 6.3 displays the total money-in investments by type and year for those firms from 1977 to 2017.

Total money-in investments over the entire period were $22.1 billion. Following the growth in new start-ups overall (see figure 6.2), investments exhibited a sharp increase starting in 2010 and reached an annual maximum of $3.2 billion in 2017 (figure 6.4). We can be confident that this maximum would be exceeded in 2018 were these data not truncated, as industry reports indicate investments in 2018 significantly exceeded those in 2017 (AgFunder 2019). Early- and late-stage VC (totaling $8.1 and $8.4 billion, respectively) represent most of the money raised by these start-ups. Even though absolute amounts increased substantially over time, the composition of investments between early-stage and late-stage VC remained quite stable. Debt financing of these firms totaled $4.2 billion but appears more sporadic, coming in large tranches when it does occur.

The ultimate fates of the 1,584 start-up companies with any associated data on transactions can be roughly divided into three types of outcomes. First, some start-ups go through a successful financial exit. In that transaction, the initial venture investors are able to exit their ownership of otherwise illiquid equity shares and realize a return on their investment. Successful exits include IPOs, M&As by other companies, and other less common buyouts, such as management buyouts or private equity buyouts. Second, the fate of start-ups that are not successful is the closure of the company—with some filing for bankruptcy, some liquidating assets, and some quietly winding down operations until they are effectively defunct. The third fate, if neither of the other two has occurred, is for a start-up to remain in business as a privately held company.

Cumulatively, for the 1,584 start-ups for which we have transaction data, we find that 150 (9.5 percent) exited via IPO, 739 (46.7 percent) exited via M&A, and 159 (10.0 percent) exited via some other buyout. Interestingly, just 49 of the start-ups (3.1 percent) reported closure or bankruptcy, implying that 487 (30.1 percent) are still in business. Not only does this ratio seem unrealistic, but others have identified a strong bias against negative news—including firm closures, small investments, and other indicators of underperformance—in VC data sets such as these (Hall and Woodward 2010). We find that of the 49 start-ups that do report closures, 90 percent of these closed within four years of their last money-in deal. It stands to reason that companies relying on VC need to raise new money every two to four years, and if they stop doing so, it is a strong indication that they have closed. Given that many of the 487 (30.1 percent) start-ups deemed "still in business" had fallen silent, lacking any newly announced deals for more than four years, we estimate that 417 (26.3 percent) face a similar probability of having closed, and thus, just 70 (4.4 percent) of the total sample were likely still in business.

Looking at exits and closures over time (figure 6.7a), we see that they occurred only sporadically prior to the mid-2000s. The number of exits began to grow steadily after 2005 and peaked in 2015. The number of closures (reported and estimated) began to increase about five years later, around 2010. Exit amounts are much more sporadic and took off dramatically in 2008, when over $2.3 billion was realized by investors (figure 6.7b). The maximum year for exit amounts was 2013, at close to $6 billion, mostly due to M&As. While the counts of exits (figure 6.7a) have displayed a smoother year-on-year growth trend, the sporadic nature of the values of exits (figure 6.7b) belies the tendency for exit valuations of start-ups in VC portfolios to be highly skewed, which has been generally observed in venture investing for decades (Gompers and Lerner 2004; Metrick 2007).

6.4 Analysis of Factors Associated with Increased Venture Capital Investments in Agriculture

There are a number of possible explanations for the sharp increase in agricultural technology start-ups beginning in 2009 (figure 6.2) and the steep rise in private investments into those companies starting in 2010 (figure 6.4). The simplest hypothesis, following the logic of Dixit and Pindyck (1994), is that prices across the industry pushed potential returns above a critical threshold. Agricultural commodity prices, indeed, increased strongly in 2007 and 2008 and then, after a correction in 2009, surged to even higher levels from 2011 to 2014 (figure 6.5a). While certainly logical, agricultural commodity prices alone do not seem sufficient to explain why VC investments began to flow into agriculture.

A more nuanced hypothesis is that the ratio of commodity prices in agriculture to prices in other sectors of the economy, particularly energy, may have diverted investments toward agriculture. And the timing of those shifts may also have played a role. "Cleantech" investment funds—which had focused primarily on the energy sector, presumably encouraged by high energy prices—may have discovered agriculture when investing in biofuels. Crude oil prices faced a sharp increase in 2007 and 2008, followed by a marked downturn in 2009, and while oil rebounded and remained around \$100/barrel from 2011 to 2014, it fell back to less than \$50/barrel within a year (figure 6.5b). At key points when oil prices dropped, investors in cleantech may have pivoted toward opportunities in agriculture as agricultural commodity prices remained higher. While such conditions seem to have held for only short windows of time (comparing figures 6.5a and 6.5b), the swings in price ratios may have been enough for venture investors to have discovered the agricultural sector. Once found, agriculture continued to be a focus of investor attention.

There is also very likely a supply-side factor, given that overall VC investments in the economy increased steadily during this period. Ewens, Nanda, and Rhodes-Kropf (2018) document an increase in investment volume by existing VC firms as well as an increase in entry by new financial intermediaries after 2006. Figure 6.6 shows that growth in investment in agtech appears to be correlated with total VC investment in the United States (PwC 2019) Thus an additional hypothesis is that a greater supply of VC coupled with lower costs of early-stage investing in this time period pushed VC investing into adjacent industries from its traditional core of software, computer/networking equipment, online businesses, and biotechnology (Ewens, Nanda, and Rhodes-Kropf 2018).

Finally, it is reasonable to expect that market signals from successful exits may have played a role. The most desired outcomes for VC investors are IPO

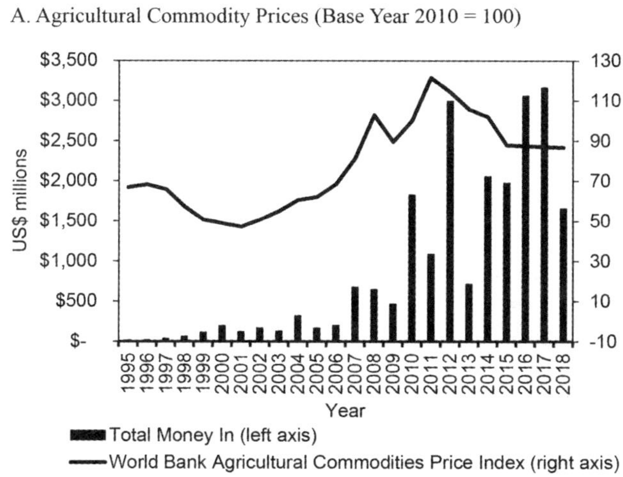

Fig. 6.5a Investments in agricultural technology start-ups plotted against agricultural commodity prices (base year 2010 = 100)

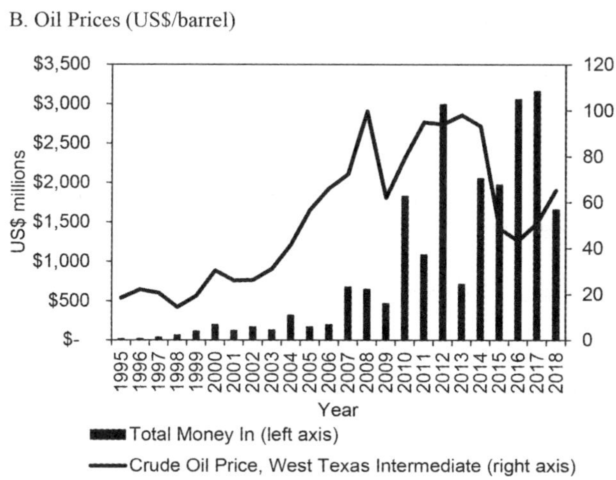

Fig. 6.5b Investments in agricultural technology start-ups plotted against oil prices (US$/barrel)

or M&A exits, as these generate the largest payoffs. Other exit outcomes, such as management buyouts or asset acquisitions, might merely return the initial investment via the sale of the start-up's assets. Gompers and Lerner (2004), Jeng and Wells (2000), and a literature spawned by such studies present evidence that successful exits influence subsequent VC investments.

Anecdotally, there were several large exits from agricultural start-ups in the years around the upturn in venture investment—including the $283 mil-

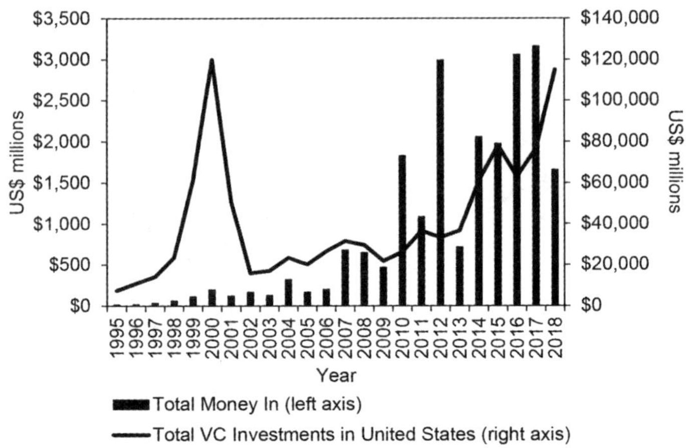

Fig. 6.6 Investments in agricultural technology start-ups plotted against total annual VC investments in the United States

lion IPO of Agria in 2007, the €1.9 billion private equity buyout of Arysta LifeScience in 2008, and the $279 million IPO of Digital Globe in 2009. According to the data, a regular rhythm of IPOs and M&As began in 2006, with significant returns first logged in 2008 (figures 6.7a and 6.7b) coinciding with the sharp increase in the numbers of new start-ups (figure 6.2) and investments (figure 6.4). In agriculture, it appears that M&As have generated much larger gross returns for VC investors than IPOs (figure 6.7b). These patterns corroborate the idea that the occurrence of IPOs and M&As signal returns being made on venture investments in agriculture, thus helping attract new investors to the newer start-ups in agriculture.

6.4.1 Regression Analysis of Investments at a Firm Level

A regression analysis was undertaken to offer a more systematic description of the relationships between VC investments in agricultural start-ups and several of these factors hypothesized to influence decisions by venture capitalists to invest. A simple framework used for analysis at the firm level is described by equation (1):

$$(1) \quad y_{it} = \alpha + \beta_1 P_{1,t-k} + \beta_2 P_{2,t-k} + \beta_3 \mathrm{VC}_t + \delta_1 ee^{m\&a}_{t-k} + \delta_2 ee^{ipo}_{t-k} + X_i\theta + u_{it},$$

where the dependent variable, y_{it}, is the sum of reported amounts of investments received by a start-up i in year t. If a start-up did not exist in year t, the observation is dropped. If a start-up did exist in year t but simply received no investment, the observation is kept, and $y_{it} = 0$. If a start-up received more than one investment in a given year, then those investments are summed.

Of the independent variables in equation (1), the $P_{1,t-k}$ are agricultural commodity prices, lagged k years, for which World Bank and Food and

A. Counts of successful exits and closures for the 1,584 startups with associated deals data, 1977-2018

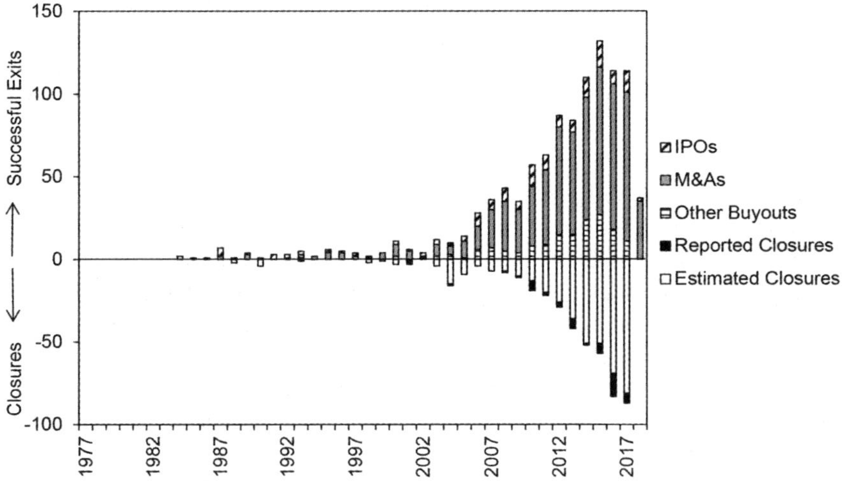

B. Amounts realized from successful exits, 1977-2018

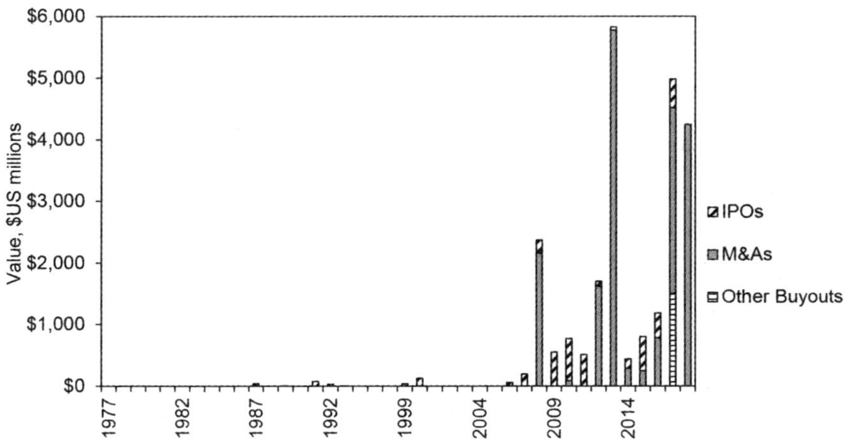

Fig. 6.7 Exits by VC investors from start-ups in agriculture

Agriculture Organization (FAO) agricultural commodity price indices as well as nominal soybean prices are considered. We also focus our analysis on possible changes in the relationship with agricultural commodity prices in the period after 2000, when they began to grow and became more volatile, by interacting agricultural prices with a date range dummy variable. The P_{2t} are nominal oil prices. The VC_t are total annual VC investments in the United States according to PricewaterhouseCoopers (PwC). The exit variables $ee_{t-k}^{m\&a}$ and ee_{t-k}^{ipo} measure the annual sum of money raised in IPO and

M&A exits of the agricultural start-ups in the sample k years prior to year t. The X_i are control variables. The sample is very heterogenous, including firms in different stages of the start-up life cycle, in different countries, and in different industry categories. Company age is used to control for stage in the start-up life cycle. Dummy variables are added for start-up locations in the United States and Europe. And dummy variables are added for the 12 industry categories described in table 6.1. Finally, u_{it} is the random error term clustered at the firm level.

Since data for the independent variables of oil prices and total VC investments were available only after 1995, the sample incorporates investment activity from 1995 to 2017. We build an unbalanced panel that consists of 12,094 observations involving 1,447 start-ups. Most of the firms were founded after 1995, with the frequency of firms entering the data set increasing toward the end of the period, according to the trend illustrated in figure 6.2. We have 2,439 firm-year observations with positive investment values and 9,655 with zero values. Due to this censoring from below, we use a tobit regression model.

We are not attempting to deal here with three important econometric challenges in working with these data. First, we are not dealing with unreported data, at two levels, in the dependent variable: for many start-ups, we observe that investments occurred, but their value is not reported, which therefore gets represented as a zero value; but we also know that there are many more investments that are entirely unreported. Second, we are not dealing with the unbalanced nature of the panel. And third, we are not attempting to deal with the endogeneity or the dynamic nature of these investments. Clearly, the trends we have summarized in the previous section are all largely moving in the same direction, making identification problematic yet beyond the scope of this chapter's objective as a descriptive exercise.

Table 6.2 presents results for the firm-level regression described by equation (1). The estimation results corroborate observations from the summary statistics displayed in figure 6.5a that trends in agricultural commodity prices are positively associated with trends in investments in agtech start-ups. The result that investments are more strongly associated with agricultural commodity prices after 2000 is consistent with the notion that growing commodity prices could have shifted the attention of VC investors toward agricultural markets.[7] Oil prices, in contrast, are negatively related to investments. This may be picking up the trends visible in figure 6.5b—namely, that investments initially remained low as oil prices increased and then later boomed as oil prices fell.[8]

7. Estimation results were found to be robust across various versions of the model that used different prices and lags, not reported here.

8. In addition, we tested the effect of the ratio of agricultural commodity prices to oil prices in regressions not reported here, with a larger ratio indicating a potentially greater return in agriculture compared to energy. We find a strong positive effect of the price ratio on the size of

Table 6.2 **Firm-level tobit regression of commodity prices, lagged exits, and total VC supply on investments made in existing firms annually over the period 1995–2017**

	Dependent variable
Independent variables	Amount invested in firm *i* in year *t*
Ag commodity prices	0.01170
	(0.02122)
Ag commodity prices after 2000	0.08949***
	(0.02211)
Oil prices	−0.31869***
	(0.08935)
Ag IPO amounts, lagged 1 year	0.01204***
	(0.00443)
Ag M&A amounts, lagged 1 year	0.00179***
	(0.00049)
Total VC invested in US	0.00026***
	(0.00005)
Firm age	−0.88796***
	(0.18550)
EU dummy	−1.99442
	(1.32819)
US dummy	9.52096***
	(2.35058)
Industry category dummies	
Constant	−85.62672***
	(18.55865)
Observations	12,094

Note: Standard errors in parentheses. *** for 1 percent significance, ** for 5 percent, and * for 10 percent. All lagged variables are lagged only one period. IPOs and M&A values for agriculture and total VC for the United States are in US$ million; ag commodity price is the nominal US soy price in US$/metric tons; oil price is West Texas Intermediate (WTI), Cushing, Oklahoma, US$ per barrel, annual, not seasonally adjusted, available at FRED. Dependent variable of annual firm deals value is in US$ million.

The variable of total VC investments reflects the overall health of VC markets and implies a greater availability of VC investments, which is plausibly associated with increased investments in start-ups in agriculture, all else being equal.[9]

Coefficients on the agIPO and agM&A variables indicate that the higher amounts realized in a previous year's exits by agtech start-ups are positively associated with higher investments in agtech start-ups in subsequent

investments when limited to the 2000–2017 window. The coefficient on the price ratio over the entire time frame is, however, not significant.

9. We also separately added an interaction between the total VC investments variable and a 2000–2018 dummy, but the resulting coefficient indicates no stronger relationship during this more limited period.

years.[10] IPOs appear to be more strongly associated than are M&As, but both are statistically significant in this regression. Both types of exits could be interpreted as playing a role in signaling returns and attracting investments into agriculture, in line with previous observations in the literature (Gompers and Lerner 2004; Jeng and Wells 2000).

Additional insights arise from the control variables in equation (1) and table 6.2. It appears that location is an important differentiator. Even though similar numbers of start-ups are found in the United States and Europe (figure 6.3b), the estimated coefficient on the US dummy variable is strongly positive and significant, while the estimated coefficient on the European dummy variable is negative and significant. This corroborates common observations that VC finance is more mature and active in the United States, and generally, US start-ups tend to receive greater VC investments compared to non-US start-ups. The estimated coefficient on the company age variable would be expected to be positive to indicate a positive relationship between age and investments: companies that have been in the market longer and have grown larger tend to receive larger VC investments, which is by design in most VC investment strategies (Gompers and Lerner 2004; Metrick 2007). The negative coefficient on company age likely reflects a high frequency of zero annual investments for older start-ups. This could arise because we still give four years of zero investments after the last reported investment to those companies that we estimate are ultimately closed; perhaps this is too generous, if many of these companies in fact closed sooner (and thus those observations should have been dropped rather than assigned a $y_{it} = 0$). Coefficients on industry category dummies are positive and significant (in order of magnitude) for (1) biotechnology, genetics, and health; (2) chemicals; and (3) software, data, and information technologies, indicating relatively more and/or larger investments are received by companies in these categories.

6.4.2 Regression Analysis of Investments at an Industry Level

To explore venture investments made at the level of the industry categories described in table 6.1 (and detailed in the appendix), a similar model is estimated:

$$(2) \quad y_{ct} = \alpha + \beta_1 P_{1,t-k} + \beta_2 P_{2,t-k} + \beta_3 \mathrm{VC}_t + \delta_1 ee^{m\&a}_{t-k} + \delta_2 ee^{ipo}_{t-k} + X_c\theta + u_{ct},$$

where the dependent variable now is y_{ct}, the annual sum of investments for all start-ups in industry category c during year t. There are reasons to believe that factors affecting investment may vary across industry or technology. As already noted, about a quarter of the start-ups in the sample are categorized in more than one industry classification due to the multidisciplinary nature of the technologies being developed or due to integration across markets. We therefore split these start-ups' investment amounts across the relevant

10. Test regressions find that exits in the same year are not significantly related to investments.

Table 6.3 **Industry-level tobit regression of commodity prices, lagged exits, and total VC supply on investments made in 12 industry categories annually over the period 1995–2017**

Independent variables	Dependent variable Amount invested in industry category *c* in year *t*
Ag commodity prices	0.29046
	(0.25682)
Ag commodity prices after 2000	0.49434**
	(0.22219)
Oil prices	−2.72660
	(1.73527)
Ag IPO amounts, lagged 1 year	0.18959***
	(0.06578)
Ag M&A amounts, lagged 1 year	0.01305*
	(0.00693)
Total VC invested in US	0.00085**
	(0.00040)
Industry category dummies	
Constant	−372.03315***
	(103.07161)
Observations	288

Note: Standard errors in parentheses. *** for 1 percent significance, ** for 5 percent and * for 10 percent. All lagged variables are lagged only one period. IPO and M&A values for agriculture and total VC for the United States are in US$ million; ag commodity price is the nominal US soy price in US$/metric tons; oil price is West Texas Intermediate (WTI), Cushing, Oklahoma, US$ per barrel, annual, not seasonally adjusted available at FRED. Dependent variable of annual firm deals value is in US$ million.

categories (e.g., if start-up A appears in two categories, we multiply its investments in year *t* by 0.5 and allocate to both categories for year *t*). The independent variables are the same as defined for equation (1).

We build a balanced panel of 12 industry categories over 24 years from 1995 to 2018 to estimate a fixed effects model also using a tobit regression model. Table 6.3 presents estimation results for investments aggregated by industry category. At this level of aggregation, coefficients on the independent variables are naturally greater than in the firm-level analysis. Agricultural commodity prices, at least after 2000, exhibit a significant positive coefficient. Oil prices are again negatively related to investments in agricultural start-ups. Coefficients on the agIPO and agM&A variables are again positive and significant, with the magnitude of the IPO coefficient again larger than the M&A coefficient. The variable for total annual VC investments is positively and significantly related to VC investments in agriculture.

At the industry level of aggregation, not all of the control variables used in the firm-level analysis—such as age or location—are meaningful. Fixed effects for industry categories are jointly statistically significant. A few of these have larger values, including (in order) (1) online services and content; (2) software, data, and information technologies; (3) marketing, processing, and distribution; and (4) agricultural input distribution and sales, indicating that relatively more frequent and/or larger aggregate investments are received in these categories.

6.5 Summary and Conclusions

The VC-backed start-up is a mechanism to contain the financial risks of prospecting and thereby manage the technical and market uncertainties of innovation. The population of start-ups developing innovations for agriculture has increased substantially over the last decade in not only high-income countries but also emerging and developing countries. Venture investments in such start-ups have grown as well—almost half as much as the estimated amounts of global corporate agriculture R&D expenditures. This first look has introduced extensive representative data on start-up companies related to agriculture and their financial transactions, and it has explored several factors that are likely to be related to the observed increase in private venture investment in agricultural R&D.

Simple tests of several hypotheses suggest that agricultural commodity prices and successful exits have been closely associated with increased VC investments in agriculture. Especially the run-up in agricultural commodity prices after 2000 appears correlated with investment levels. Both IPO and M&A exit amounts realized by agricultural start-ups are associated with subsequent investments at both the firm and industry levels. IPOs appear to have a stronger relationship with new investments than do M&As, even though a much larger share of the returns realized from exits come from M&As. Investments in agricultural start-ups are to some extent technology specific, favoring biotech, online businesses, software, commodity processing, and agricultural input dealers. There is also evidence that start-ups in the United States receive more venture investments than start-ups in other countries, all else being equal.

This analysis sheds light on an important new source of R&D expenditure that has the potential to transform many aspects of private R&D for agriculture, altering the risk profile of innovations being pursued, the networks of highly skilled human capital being accessed, and the market power of companies introducing innovations throughout the agricultural value chain. Much is needed in the way of further economic analysis of these trends to improve on current models and explore the factors potentially driving such investments (e.g., public sector research, other sources of technological opportunity, increased labor costs, or shifts in consumer demand) and the determinants and impacts of different types of exits (with IPOs creating independent competitors but M&As putting new technologies under the control of industry incumbents). Venture capital has discovered agriculture, but it has only begun to impact agriculture.

Appendix

Characterization of Agricultural Start-Up Firms by Industry and Technology

In the data obtained from commercial vendors, either start-up companies self-report or the commercial data vendor composes a business description, usually a short paragraph, and assigns an industry code or segment categorization. These vary, however, across vendors. In order to ascertain more uniformly the industry or technology in which start-ups are engaged, we queried and filtered the company description and industry categorization fields to assign start-ups to 12 categories, as summarized in table 6.1. These categories are relied on to introduce field-specific controls in our estimations (tables 6.2 and 6.3). The following notes describe in greater detail the types of businesses that are included in each of the categories:

Business and financial services

1. Real estate, land brokerages
2. Human resource management, labor contracting, training and education services
3. Financial services, investment
4. Insurance, risk management
5. Industry associations and advocacy
6. Economic development and regional development organizations
7. Business-to-business (B2B) services or marketplaces (in combination with the online category)
8. Publishing, catalogs, information for industry clients (may be in combination with the online category)
9. Consulting and advisory services
10. Contract research services

Online services and content

1. Online, website, web, or portal; often platform
2. B2B or business-to-consumer (B2C); almost always in combination with another appropriate category
3. Apps or mobile; often in combination with the software, data, and information technologies category

Genetics, breeding, biotech, and health

1. Companies described as biotech
2. Companies that mention genetics

3. Breeding
4. Biological control
5. Biopesticides
6. Biofertilizers, compost, biochar, other biologically derived soil amendments
7. Microbial/microbiome
8. Animal health, including vaccines (but *not* feed additives)
9. Animal reproduction, such as sexing, artificial insemination (AI)

Chemicals

1. Agricultural chemical manufacturing
2. Any of the chemical "-cides" (pesticides, insecticides, herbicides, fungicides, etc.), if not explicitly biological (i.e., *not* biopesticides or *not if* explicitly described as a protein or peptide, which are instead included in the "genetics, breeding, biotech, and health" category)
3. Mention of a specific class of chemical compound that characterizes the company's products
4. Inert materials with beneficial properties as soil additives, fillers, growth media, weed blockers, mulches, and so on
5. Nanomaterials

   Note: The use of this category indicates R&D or manufacturing, not merely distribution or "provider" of chemical products.

Electronic devices, sensors, and systems (electronic hardware)

1. Use of the words *device* or *sensor*, smart or automated systems, measurement or monitoring in electronics context
2. Hardware (as opposed to software)
3. Robots, drones, unmanned or autonomous vehicles (UAVs)
4. Lighting or LED systems for contained or indoor agriculture
5. Control systems

   Note: This category includes technologies/products that would be in electrical engineering, not mechanical, civil, or hydrological engineering (these are under the "machinery and equipment [mechanical hardware]" category).

Software, data, and information technologies

1. Software or app
2. Data
3. Analytics
4. Artificial intelligence or machine learning
5. Blockchain or distributed ledger

Machinery and equipment (mechanical hardware)

1. Manufacture of farm machinery or equipment
2. Development or sales of vertical or indoor agricultural equipment and infrastructure (not control systems or automation, which are

included under the "electronic devices, sensors, and systems [electronic hardware]" category)

Note: This category does not include the distribution, import, or sales of farm machinery and equipment; these are under the "agricultural input distribution and sales" category.

Agricultural input distribution and sales

1. Distribution, sales, retail, wholesale, supply, provision (but *not* manufacturing) of a range of agricultural inputs, including
   a. seeds, plant starts
   b. agricultural chemicals, pesticides, fertilizers
   c. biological amendments, inputs
   d. animal feed, feed additives, supplements
   e. animal health, veterinary products and supplies
   f. young live animals (chicks, fish fry, etc.)
   g. farm supplies, aquaculture supplies
   h. machinery and equipment (for farms, ranches, aquaculture, fishing, timber operations)
   i. parts and services
2. Small minority include agricultural services, such as contract harvesting, piecework, agronomic consulting services, monitoring, and management
3. Does not include provision of or contracting of agricultural labor; human resource services are all under the "business and financial services" category
4. When the input is animal feed (1.d above), this category is often used in combination with the agricultural outputs marketing, processing, and manufacturing category if the company also manufactures or produces the animal feed, which often involves grain or oilseed milling

Agricultural production and farming

1. Actual operation of a farm or other agricultural production operation such as a ranch or fish hatchery
2. Cultivation
3. Production
4. Often includes the phrase *provision of agricultural services*
5. Often mentions actual commodities produced
6. In combination with the agricultural outputs marketing, processing, and manufacturing category if the company is a vertically integrated agribusiness (e.g., in livestock, oil palm)
7. In combination with the agricultural outputs marketing, processing, and manufacturing category if vertically integrated fresh market (e.g., fruit, vegetable, produce)
8. In combination with the "agricultural output marketing, processing,

and manufacturing" category *and* with the "consumer products or services" category if it includes the phrases *community-supported agriculture* (CSA), *farm to table*, *locally produced*, and so on

Agricultural output marketing, processing, and manufacturing

1. Postharvest marketing, distribution, export/import, brokering
2. Transportation, logistics
3. Processing, milling
    a. animal slaughter, meat processing, meat packing
    b. grain milling, feed manufacturing
    c. oilseed pressing, processing
    d. cotton ginning
    e. sawmills
    f. ethanol plants
4. Other fermentation, extraction, separation, purification for ingredient manufacturing; animal feed additives (amino acids, micronutrients, etc.)
5. Food manufacturing; food brand or category for broad market (i.e., national or commodity-wide)
6. Wineries, breweries, distilleries
7. Farmers' markets, local food marketing

Consumer products and services

1. Explicit mention of the consumer, home, or household
2. Retail
3. A specific final product, often branded
4. Direct marketing or distribution to final consumer (not to stores, restaurants, or other food services)
5. Consumer connected to production/distribution (e.g., community agriculture, farm to table, farm share schemes)
6. Mention of garden, gardening supplies, garden equipment, indoor gardening systems, if clearly intended for home (not for horticulture or greenhouse industry)

Unspecified

1. Unable to determine: combined industry/technology descriptions are too general or missing altogether

References

AgFunder. 2015. *AgTech Investing Report—2014*. San Francisco: AgFunder.

———. 2018a. *China AgriFood Startup Investing Report*. San Francisco: AgFunder.

———. 2018b. *India AgriFood Startup Investing Report—5-Year Review 2013–2017*. San Francisco: AgFunder.

———. 2019. *AgriFood Tech Investing Report—2018*. San Francisco: AgFunder.
Alston, Julian M., Matthew A. Andersen, Jennifer S. James, and Philip G. Pardey. 2010. *Persistence Pays: U.S. Agricultural Productivity Growth and the Benefits from Public R&D Spending*. New York: Springer.
Arqué-Castells, Pere. 2012. "How Venture Capitalists Spur Invention in Spain: Evidence from Patent Trajectories." *Research Policy* 41:897–912.
Audretsch, David B. 1994. "Business Survival and the Decision to Exit." *Journal of the Economics of Business* 1 (1): 125–37.
———. 1995. "The Propensity to Exit and Innovation." *Review of Industrial Organization* 10 (5): 589–605.
Baum, Joel A. C., and Brian S. Silverman. 2004. "Picking Winners or Building Them? Alliance, Intellectual, and Human Capital as Selection Criteria in Venture Financing and Performance of Biotechnology Startups." *Journal of Business Venturing* 19:411–36.
Carey, Janis M., and David Zilberman. 2002. "A Model of Investment under Uncertainty: Modern Irrigation Technology and Emerging Markets in Water." *American Journal of Agricultural Economics* 84 (1): 171–83.
CBInsights. 2017. *Ag Tech Heats Up: 5 Trends Shaping the Future of Farming & Agribusiness*. December 12, 2017.
Cumming, D., and S. Johan. 2017. "The Problems with and Promise of Entrepreneurial Finance." *Strategic Entrepreneurship Journal* 11 (3): 357–70.
Da Rin, Marco, and Maria Fabiana Penas. 2017. "Venture Capital and Innovation Strategies." *Industrial and Corporate Change* 26 (5): 781–800.
Deloof, Marc, and Tom Vanacker. 2018. "The Recent Financial Crisis, Startup Financing and Survival." *Journal of Business Finance & Accounting* 45 (7–8): 928–51.
Dias, Cleidson N., Francisco Jardim, and Luiz O. Sakuda. 2019. *Radar AgTech Brasil 2019: Mapeamento das Startups do Setor Agro Brasileiro*. São Paulo, Brazil: EMBRAPA, SP Ventures, and Homo Ludens. http://www.radaragtech.com.br.
Dixit, Avinash K., and Robert S. Pindyck. 1994. *Investment under Uncertainty*. Princeton, NJ: Princeton University Press.
Dutia, Suren G. 2014. *Agtech: Challenges and Opportunities for Sustainable Growth*. Kansas City, MO: Ewing Marion Kauffman Foundation.
Engel, Dirk, and Max Keilbach. 2007. "Firm-Level Implications of Early Stage Venture Capital Investment—an Empirical Investigation." *Journal of Empirical Finance* 14:150–67.
Ewens, Michael, Ramana Nanda, and Matthew Rhodes-Kropf. 2018. "Cost of Experimentation and the Evolution of Venture Capital." *Journal of Financial Economics* 128 (3): 422–42.
Félix, Elisabete Gomes Santana, Cesaltina Pacheco Pires, and Mohamed Azzim Gulamhussen. 2013. "The Determinants of Venture Capital in Europe—Evidence across Countries." *Journal of Financial Services Research* 44 (3): 259–79.
Finistere Ventures. 2019. *2018 Agtech Investment Review*. San Diego: Finistere Ventures.
Food and Agricultural Organization (FAO). 2018. *The State of Food Security and Nutrition in the World 2018: Building Climate Resilience for Food Security and Nutrition*. Rome: Food and Agricultural Organization.
Fuglie, Keith. 2016. "The Growing Role of the Private Sector in Agricultural Research and Development World-Wide." *Global Food Security* 10:29–38.
Fuglie, Keith, Paul Heisey, John King, Carl E. Pray, and David Schimmelpfennig. 2012. "The Contribution of Private Industry to Agricultural Innovation." *Science* 338 (6110): 1031–32.
Fuglie, Keith O., Paul W. Heisey, John L. King, Carl E. Pray, Kelly Day-Rubenstein,

David Schimmelpfennig, Sun Ling Wang, and Rupa Karmarkar-Deshmukh. 2011. *Research Investments and Market Structure in the Food Processing, Agricultural Input, and Biofuel Industries Worldwide*. ERR-130, US Department of Agriculture, Economics Research Service.

Gaulé, Patrick. 2015. "Patents and the Success of Venture-Capital Backed Startups: Using Examiner Assignment to Estimate Causal Effects." Center for Economic Research and Graduate Education-Economics Institute (CERGE-EI), Working Paper Series 546, Charles University, Prague, Czech Republic.

Giovannetti, Giorgia, Giorgio Ricchiuti, and Margherita Velucchi. 2011. "Size, Innovation and Internationalization: A Survival Analysis of Italian Firms." *Applied Economics* 43 (12): 1511–20.

Gompers, Paul Alan, and Josh Lerner. 2004. *The Venture Capital Cycle*. Cambridge, MA: MIT Press.

Gooch, Elizabeth, and Fred Gale. 2018. "China's Foreign Agriculture Investments." Economic Information Bulleting No. 192. US Department of Agriculture, Economic Research Service.

Graff, Gregory, Annabelle Berklund, and Kathay Rennels. 2014. *The Emergence of an Innovation Cluster in the Agricultural Value Chain along Colorado's Front Range*. Fort Collins: Colorado State University Press.

Graff, Gregory, Gordon Rausser, and Arthur Small. 2003. "Agricultural Biotechnology's Complementary Intellectual Assets." *Review of Economics and Statistics* 85 (2): 349–63.

Groh, Alexander Peter, and Johannes Wallmeroth. 2016. "Determinants of Venture Capital Investments in Emerging Markets." *Emerging Markets Review* 29:104–32.

Hall, Robert E., and Susan E. Woodward. 2010. "The Burden of the Nondiversifiable Risk of Entrepreneurship." *American Economic Review* 100:1163–94.

Holmes, Phil, A. Hunt, and Ian Stone. 2010. "An Analysis of New Firm Survival Using a Hazard Function." *Applied Economics* 42 (2): 185–95.

Hu, Ruifa, Qin Liang, Carl Pray, Jikun Huang, and Yanhong Jin. 2011. "Privatization, Public R&D Policy, and Private R&D Investment in China's Agriculture." *Journal of Agricultural and Resource Economics* 36 (2): 416–32.

Huffman, Wallace E., and Robert E. Evenson. 2006. *Science for Agriculture: A Long-Term Perspective*. Oxford: Blackwell.

Jeng, Leslie A., and Philippe C. Wells. 2000. "The Determinants of Venture Capital Funding: Evidence across Countries." *Journal of Corporate Finance* 6 (3): 241–89.

Kolympiris, C., S. Hoenen, and N. Kalaitzandonakes. 2017. "Geographic Distance between Venture Capitalists and Target Firms and the Value of Quality Signals." *Industrial and Corporate Change* 27 (1): 189–220.

Kolympiris, Christos, and Nicholas Kalaitzandonakes. 2013. "The Geographic Extent of Venture Capital Externalities on Innovation." *Venture Capital* 15 (3): 199–236.

Kolympiris, Christos, Nicholas Kalaitzandonakes, and Douglas Miller. 2015. "Location Choice of Academic Entrepreneurs: Evidence from the US Biotechnology Industry." *Journal of Business Venturing* 30 (2): 227–54.

Kortum, Samuel, and Josh Lerner. 2000. "Assessing the Contribution of Venture Capital to Innovation." *RAND Journal of Economics* 31 (4): 674–92.

KPMG. 2018. *Venture Capital Investment in Agtech Continues to Soar*. KPMG Insights, October 30, 2018. https://home.kpmg/au/en/home/insights/2018/08/venture-capital-investment-agtech.html.

Metrick, Andrew. 2007. *Venture Capital and the Finance of Innovation*. Hoboken, NJ: John Wiley & Sons.

Mondin, Mateus, and José Tomé. 2018. *2º Censo AgTech Startups Brasil*. Piracicaba,

Brazil: Escola Superior de Agricultura Luiz de Queiroz (ESALQ), University of São Paulo and AgTechGarage. https://www.agtechgarage.com/censo/.

Nadeau, Pierre. 2011. "Innovation and Venture Capital Exit Performance." *Strategic Change* 20 (7–8): 233–52.

Pardey, Philip G., Julian Alston, and Vernon Ruttan. 2010. "The Economics of Innovation and Technical Change in Agriculture." In *Handbook of Economics of Innovation*, vol. 2, edited by Bronwyn H. Hall and Nathan Rosenberg, 939–84. Amsterdam: Elsevier.

Pardey, Philip G., and Nienke M. Beintema. 2001. *Slow Magic: Agricultural R&D a Century after Mendel*. Washington, DC: Agricultural Science and Technology Indicators Initiative (ASTI), International Food Policy Research Institute (IFPRI).

Pardey, Philip G., Nienke M. Beintema, Steven Dehmer, and Steven Wood. 2006. *Agricultural Research: A Growing Global Divide?* Washington, DC: Agricultural Science and Technology Indicators Initiative (ASTI), International Food Policy Research Institute (IFPRI).

Pardey, Philip G., Connie Chan-Kang, Jason M. Beddow, and Steven P. Dehmer. 2015. "Long-Run and Global R&D Funding Trajectories: The U.S. Farm Bill in a Changing Context." *American Journal of Agricultural Economics* 97 (5): 1312–23.

Pardey, Philip G., Connie Chan-Kang, Steven P. Dehmer, and Jason M. Beddow. 2016. "Agricultural R&D Is on the Move." *Nature News* 537 (7620): 301.

Pe'er, Aviad, and Thomas Keil. 2013. "Are All Startups Affected Similarly by Clusters? Agglomeration, Competition, Firm Heterogeneity, and Survival." *Journal of Business Venturing* 28 (3): 354–72.

PitchBook. n.d. "What Are Industry Verticals?" Accessed February 16, 2021. https://pitchbook.com/what-are-industry-verticals.

PricewaterhouseCoopers (PwC). 2019. "MoneyTree Explorer." Accessed February 16, 2021. https://www.pwc.com/us/en/industries/technology/moneytree/explorer.html#/.

Puri, Manju, and Rebecca Zarutskie. 2012. "On the Life Cycle Dynamics of Venture-Capital and Non-venture-capital-financed Firms." *Journal of Finance* 67 (6): 2247–93.

Rausser, Gordon, Ben Gordon, and James Davis. 2018. "Recent Developments in the California Food and Agricultural Technology Landscape." *ARE Update* 21 (4): 5–8. https://giannini.ucop.edu/publications/are-update/issues/2018/21/4/recent-developments-in-the-california-food-and-agr/.

Sunding, David, and David Zilberman. 2001. "The Agricultural Innovation Process: Research and Technology Adoption in a Changing Agricultural Sector." In *Handbook of Agricultural Economics*, edited by Bruce L. Gardner and Gordon C. Rausser, 207–61. Amsterdam: Elsevier.

Wright, Brian. 2009. *International Grain Reserves and Other Instruments to Address Volatility in Grain Markets*. Washington, DC: World Bank.

Zilberman, David, Liang Lu, and Thomas Reardon. 2019. "Innovation-Induced Food Supply Chain Design." *Food Policy* 83:289–97.

Comment Michael Ewens

6.C1 Summary

This chapter investigates the trends in global private equity (PE) and venture capital (VC) investments in agriculture start-ups. After finding evidence for rapid increases in capital flow to these start-ups—particularly after the financial crisis of 2008—the chapter explores the sources of changes. Specifically, the capital provided to these start-ups is growing relative to the supply of capital invested by the public sector and public firms. Next, a regression analysis confirms that investment in agricultural sectors is strongly correlated with both past liquidity events in the sector and changes in prices for major commodities. The results are consistent with investors responding to investment opportunity signals in the agricultural space.

Answers to these questions are important for researchers and policy makers who aim to support the agricultural sector and its innovation. More broadly, the analysis of changing investor behaviors, such as new allocations to new industries, reveals where start-up financing constraints lie. Finally, the results contribute to perennial debates around public versus private research and development (R&D) spending.

The comments on this chapter focus on several topics, including data construction and interpretation and suggestions for additional analyses.

6.C1.1 Combining Databases: Benefits and Pitfalls

The chapter describes a major data exercise merging three databases: Crunchbase, PitchBook (Morningstar), and VentureSource (Dow Jones). The authors should be applauded for combining these related but distinct sources of data. However, such merges face challenges when data providers differ in their coverage and industry classification methodologies. Consider first the VentureSource database provided by Dow Jones. In my experience using this data, I have learned that their best coverage is for US-based start-ups backed by VC. Informal conversations with the data provider also revealed that the data quality is high only after 1990 (the firm was founded in 1987). Next, PitchBook provides significantly wider coverage by region than VentureSource. Early focus was on the US PE ecosystem but has grown and—in my opinion—improved over the last five to seven years. Given their founding in 2007, it is not clear where their historical data was sourced, which is an important uncertainty when merging with databases that have

Michael Ewens is a professor of finance and entrepreneurship at the California Institute of Technology and a research associate of the National Bureau of Economic Research.

For acknowledgments, sources of research support, and disclosure of the author's material financial relationships, if any, please see https://www.nber.org/books-and-chapters/economics-research-and-innovation-agriculture/comment-venture-capital-and-transformation-private-rd-agriculture-first-look-ewens.

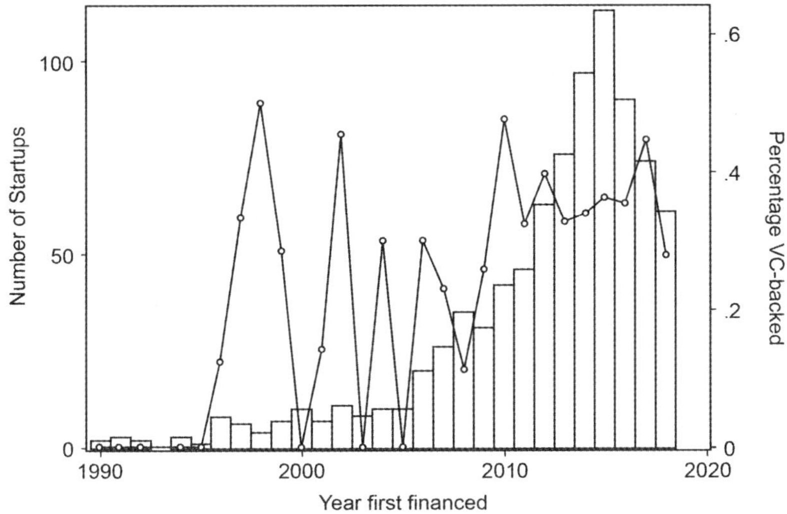

Fig. 6.C1 Agtech start-ups, PitchBook

Note: The figure reports the count of VC/PE-backed US-based agtech start-ups from 1990 to 2018. The left axis and bars report the raw counts of the number of start-ups raising such capital. The line reports the fraction that are "VC backed" according to PitchBook.

longer time series. Finally, Crunchbase fills out the financing coverage. This database is best known for its coverage of US early-stage (often pre-VC) financings. It began in 2007 by a technology blogger to keep track of start-ups covered on its site and grew as a wiki-style page. It also appears to have benefited from the switch away from PDF to XML-formatted regulatory filings for PE exemption notices in 2009. Given its short history and narrow industry focus in the early years, its quality for agriculture start-ups is unclear.

My suggestion to the authors is to first motivate the merge of these three databases. For example, is there evidence that one has poor coverage of exits or nonsoftware companies? The main concern is that the quality, coverage, and definitions differ widely across (and possibly within) data providers. Note that each of these companies likely makes most of their revenue from nonacademic customers, which means they are less concerned with historical data and have resources devoted to the current period. The best motivation for this merge would be to fill in gaps in each databases' coverage. Alternatively, the authors could pick one as a "master" data set and use the remaining two to fill in coverage gaps or missing values.

As an example of possible time-varying coverage, consider a query of PitchBook for US-based agricultural start-ups financed in 1990–2018 for "all" investor types in figure 6.C1.

Several questions emerge from this figure that warrant some discussion

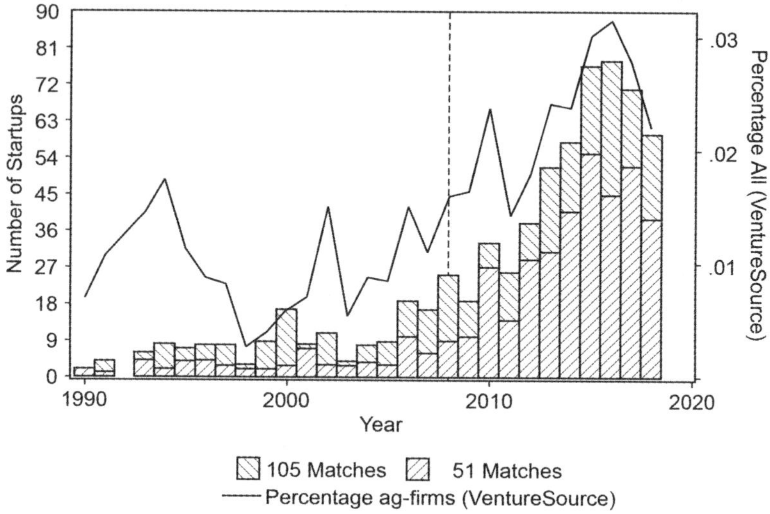

Fig. 6.C2 Merge of VentureSource and PitchBook

Note: The figure reports the counts of successful merges of start-ups in VentureSource and PitchBook using two different search queries for VentureSource.

in the chapter. First, note the increase in counts that coincides with the founding of PitchBook in 2007. Can the authors confirm that this is a real phenomenon rather than a coverage change? Second, for the chapter's analysis of VC-backing changes over the sample period, how does one explain the line of percent VC backed? It exhibits significant variation but no major break in the 2008–9 period.

Next, I was able to conduct a quick merge of VentureSource and PitchBook for US-based VC-backed agricultural start-ups. The first challenge is selecting industries. Unlike publicly traded firms that have at least one Standard Industrial Classification (SIC) code, private firms have self-assigned industries or classifications given by data providers. Private firm data providers do not always use SIC or North American Industry Classification System (NAICS) codes, which limits merging across databases. My attempt to get agricultural start-ups in VentureSource and PitchBook was thus challenging. For VentureSource, I choose the broad category "Agriculture and Forestry" with added flags in business descriptions for "farm," "harvest," and "agriculture." The latter is important because tractor guidance software is categorized as "Software," but the start-ups have "farm" in their description. I followed the chapter's approach to querying PitchBook.

The merge of VentureSource and PitchBook was done using the start-up's name (after some basic cleaning). Figure 6.C2 presents the number of observations from the successful merge of the search results using the "Agriculture and Forestry" category (51 matches) and the 105 keywords. The

General Information

Description
Developer of agricultural data platform design to offer insights and analytics for the agricultural commodities supply chain. The company's agricultural data platform is an application that converts big-data from satellites and weather stations into actionable, macro-level insights on agricultural production, enabling agricultural businesses to avail information of flooding, drought, and other adverse conditions that might affect their commodities.

Most Recent Financing Status (as of 13-Aug-2018)
The company raised $100,000 of convertible debt financing from Mindset Ventures on February 20, 2017.

Website	www.aerialintel.com		
Entity Types	Private Company	**University**	Venture Capital
Legal Name	Aerial Intelligence, Inc.	**Primary Industry**	Media and Information Services (B2B)
Business Status	Generating Revenue	**Other Industries**	Database Software
Ownership Status	Privately Held (backing)		Business/Productivity Software
Financing Status	Venture Capital-Backed	**Verticals**	Artificial Intelligence & Machine Learning
			Big Data
			Saas
			TMT

Fig. 6.C3 Example description of start-up in PitchBook not listed as "agtech"

results indicate not only that casting a wide net in any search is important for increasing the sample size but also that individual matches demand random checks for false positives.

A closer analysis of the failed merges suggests that some hand collection is necessary. Aerial Intelligence (figure 6.C3) is found in PitchBook, but not as an agriculture firm. The start-up's description strongly suggests it is agtech. Next, the start-up VinSense is not found in VentureSource. It has the following business description: "Developer of a crop management software designed to enhance crop uniformity and increase crop volume. The company's software helps to improve crop management using soil sensors and offers optimal soil nutrient management, enabling producers, field managers and winemakers to manage soil moisture, pruning, irrigation, canopy management and water conservation."[1]

An analysis of VinSense's financings shows that over half of its capital raised was in the form of government grants, while its equity investors made abnormally small investments. This example makes clear that data providers also have different methodologies to determine what constitutes "VC backed." It also shows how valuable merging different databases can be for improving coverage. I would like to see more discussion of the rules each data provider uses when classifying and collecting data for the industries of interest.

1. See the company website at https://www.vinsense.net/product-services/.

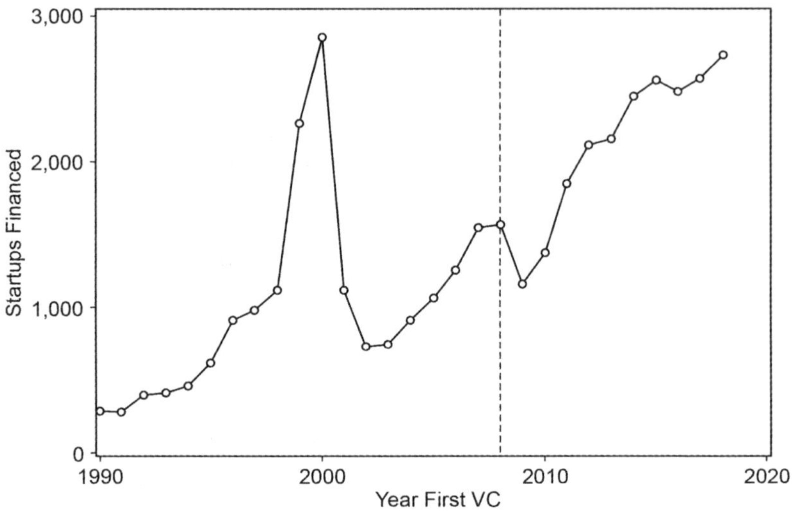

Fig. 6.C4 Number of start-ups financed by VC

Because data providers change methodology and quality over time, merging them can create spurious trends. This leads to two concrete suggestions:

1. Run the analysis on each database separately and demonstrate that results all go in the same direction.
2. Manually merge all three using a wide net on each database and document why they disagree.

6.C1.2 What's the Counterfactual?

The chapter argues that there has been a meaningful increase in capital provided to agriculture start-ups. The authors could do more to tease out the overall or macro trends from these changes. For example, figure 6.C4 shows the number of start-ups backed by VC since 1990 (according to VentureSource). The vertical dashed line shows the break in financing proposed by the authors. Clearly, the overall VC market experienced a change around 2008. Thus I suggest that the authors isolate an area *within* agriculture that grew differently. For example, one could conduct a structural break test with an unknown break in mean (constant), repeating the exercise for the changes in agricultural prices.

6.C1.3 What Are the Next Steps?

The chapter has the opportunity to explore deeper issues in both agriculture and venture capital. The chapter hints at one direction: "Several factors may have affected the hurdle rate, such as an increase in the ratio of agricultural prices to nonagricultural commodity prices, the occurrence of large exit events in highly visible start-ups, the emergence of new technological opportunities based on advances in enabling technologies (such as cheaper genome sequencing, genome editing, or data capacity of sensors and networks), and changes in (agricultural) labor markets in both high-income and middle-income countries."

One suggestion is to follow a similar strategy found in Ewens, Nanda, and Rhodes-Kropf (2018). They study the impact of changing start-up costs after the introduction of the cloud. Their focus was on the information technology (IT) sector; however, it is likely that impacts are *within* agriculture. The same technological shock could be used to study the role of investor value-add to this industry or how capital flows between different sectors in the agricultural space.

Another avenue for additional analysis begins with the premise that the increase in capital to agriculture is real. One can ask, Who are the investors? This is an interesting question because agriculture is a nontraditional space for both VC and PE. One prediction is that existing investors are pivoting toward agriculture to exploit new technology in the space. Here, changes in investing represent not a demand shock but rather a spillover from a lack of investment opportunities elsewhere. Alternatively, the new investments have new investor entrants that are VCs. Such a pattern is consistent with a supply-side shock or exits from established agtech firms. One way to investigate this issue would require tracking the work histories of the partners in the start-up financings. Finally, it is possible that the growth in agtech is facilitated by new *types* of investors (e.g., incubators, corporate VC, PE hybrids)? If so, then the facts would be consistent with existing VCs having

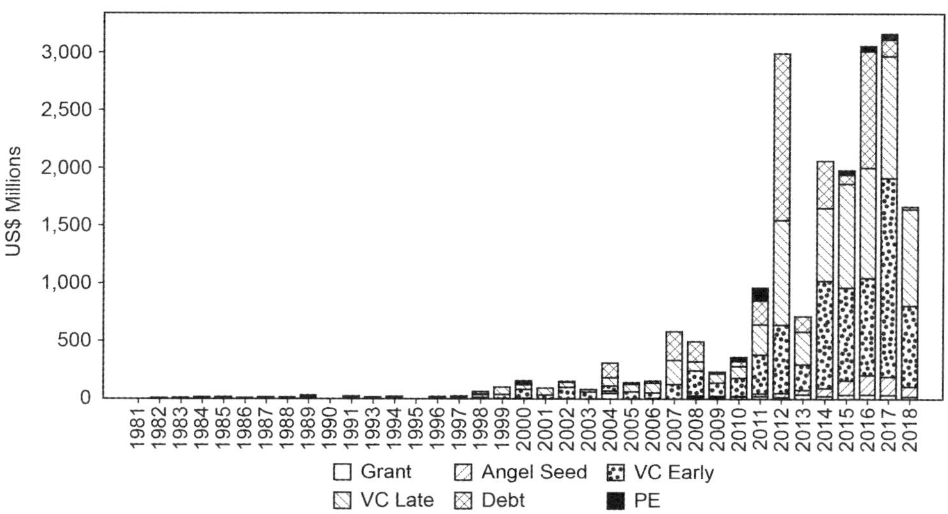

Fig. 6.C5 Figure 6.4 from chapter 6

real constraints on either skill or capital for agricultural investment opportunities. In fact, figure 6.4 in the chapter (re-created below) shows that VC is not the only source of capital in these start-ups and there is some evidence of changing composition of investors.

6.C2 Conclusion

The chapter documents changes to the entrepreneurial finance ecosystem in agriculture. Moreover, it documents strong correlations between the flows to start-ups and signals of investment opportunities. My suggestions for the authors are threefold. First, they should conduct a careful review of the database creation with particular attention paid to the variation in industry and coverage differences by data provider. Next, more evidence is needed to convincingly demonstrate that the financing environment changed in agriculture in ways different from that experienced in all of the start-up ecosystem. Finally, the authors have many opportunities to explore how VC investment dynamics are connected to changes in the agriculture industry or changes to the supply side of the market.

Contributors

Michael J. Andrews
Department of Economics
University of Maryland, Baltimore County
1000 Hilltop Circle
Baltimore, MD 21250

Bradford Barham
Department of Agricultural and Applied Economics
University of Wisconsin-Madison
427 Lorch Street
Madison, WI 53706

Nicola Bianchi
Strategy Department
Kellogg School of Management
2211 Campus Drive
Evanston, IL 60208

Matthew Clancy
Department of Economics
Iowa State University
513 Farm House Lane
Ames, IA 50011

Michael Ewens
California Institute of Technology
MC 228-77
1200 East California Avenue
Pasadena, CA 91125

Jeremy Foltz
Department of Agricultural and Applied Economics
University of Wisconsin-Madison
427 Lorch Street
Madison, WI 53706

Alberto Galasso
Rotman School of Management
University of Toronto
105 St. George Street
Toronto, ON M5S 3E6 Canada

Gregory D. Graff
Department of Agricultural and Resource Economics
Colorado State University
1200 Center Avenue Mall
Fort Collins, CO 80523

Paul Heisey
Economic Research Service
US Department of Agriculture
355 E Street SW
Washington, DC 20024-3221

Brent Hueth
Economic Research Service
805 Pennsylvania Avenue
Kansas City, MO 64105

Jared Hutchins
Department of Agricultural and Consumer Economics
University of Illinois at Urbana-Champaign
1301 W Gregory Drive
Urbana, IL 61801

Yongjie Ji
Center for Agricultural and Rural Development
Iowa State University
518 Farm House Lane
Ames, IA 50011-1054

Ana Paula Melo
Department of Agricultural and Applied Economics
University of Wisconsin-Madison
427 Lorch Street
Madison, WI 53706

Keith Meyers
D-IAS, Danish Institute for Advanced Study
University of Southern Denmark
Campusvej 55
5230 Odense M, Denmark

GianCarlo Moschini
Department of Economics
Iowa State University
518 Farm House Lane
Ames, IA 50011-1054

Petra Moser
Department of Economics
NYU Stern
44 West 4th Street
New York, NY 10012

Paul W. Rhode
Economics Department
University of Michigan
611 Tappan Street
Ann Arbor, MI 48109

Michael J. Roberts
Department of Economics
University of Hawai'i at Mānoa
2424 Maile Way
Honolulu, HI 96822

Guilherme Rosa
Department of Animal and Dairy Sciences
University of Wisconsin-Madison
1675 Observatory Drive
Madison, WI 53706

Bhaven N. Sampat
Department of Health Policy and Management
Columbia University
722 W 168th Street
New York, NY 10032

Felipe de Figueiredo Silva
Department of Agricultural Sciences
Clemson University
Clemson, SC 29634

David Zilberman
Department of Agricultural and Resource Economics
207 Giannini Hall, #3310
University of California
Berkeley, CA 94720-3310

Author Index

Subject Index

Note: Page numbers followed by "f" or "t" refer to figures or tables, respectively.